HELPING RELATIONSHIPS AND STRATEGIES

Second Edition

HELPING RELATIONSHIPS AND STRATEGIES

Second Edition

David E. Hutchins
Virginia Polytechnic Institute and State University

Claire G. Cole
Southwest Virginia Regional Assessment Center
Virginia Polytechnic Institute and State University

Brooks/Cole Publishing Company
Pacific Grove, California

Brooks/Cole Publishing Company
A Division of Wadsworth, Inc.

Printed in the United States of America
10 9 8 7 6 5 4 3 2 1

Library of Congress Cataloging-in-Publication Data

Hutchins, David E.
Helping relationships and strategies / David E. Hutchins, Claire G. Cole. — 2nd ed.
p. cm.
Includes bibliographical references and index.
ISBN 0-534-16404-8
1. Counseling. 2. Helping behavior. I. Cole, Claire G. II. Title.
BF637.C6H85 1991
158'.3—dc20 91-21798
CIP
AC

Sponsoring Editor: CLAIRE VERDUIN
Editorial Associate: GAY C. BOND
Production Editor: PENELOPE SKY
Manuscript Editor: BETTY DUNCAN
Interior and Cover Design: CLOYCE J. WALL
Art Coordinator: CLOYCE J. WALL
Interior Illustration: LOTUS ART, ANDREW MYER
Typesetting: HARRISON TYPESETTING, INC.
Printing and Binding: THE MAPLE-VAIL BOOK MANUFACTURING GROUP

To Marilyn, Mark, and John

—DAVE

To Steve, Patrick, Scott, and Robin

—CLAIRE

PREFACE

WE expect this book to be used by insightful and creative teachers, clinicians, and students who challenge themselves to find increasingly effective methods of helping people resolve problems. We have made significant changes in this edition in order to provide a practical way of integrating diverse methods.

Systematic integration of different approaches to the helping process is the most significant and challenging development in counseling and psychotherapy today (Frank, 1981; Norcross, 1986; Smith, 1982; Ward, 1983). It is significant because new methods arise from foundations laid by such major pioneers as Rogers, Ellis, Wolpe, and others. The challenge is in determining which approaches apply to which client (Paul, 1967).

The TFA Model

We use the TFA systems model (Hutchins & Vogler, 1988) throughout, defining behavior as "the integration of *t*houghts, *f*eelings, and *a*ctions" (Hutchins, 1984). In Chapter 2 we introduce the TFA triangle, a practical, easily understood graphic method of integrating numerous helping procedures and theoretical approaches. The TFA model has been tested in classrooms and clinical practice for more than a decade, and has extraordinary appeal for many people:

- *Beginning students* find the TFA model easy to understand. It helps them organize and conceptualize the many different methods and theories used in the helping process.
- *Teachers, clinicians, and advanced students* use the TFA model to integrate a wide range of techniques and strategies with diverse theoretical concepts.
- *Clients* immediately grasp the central TFA concepts. The visual stimulus of the TFA triangle often helps clients understand their behavior clearly and quickly. The critical test of any helping relationship is whether it makes a positive difference to the client. In a process that relies on the client to implement changes it is necessary for the client, as well as for the helper, to understand essential parts of the process.

Organization

The book is in four parts, which makes it accessible to those who want to learn (or refresh themselves on) basic skill development. In Part I we focus on the helping process, introduce the TFA model, and show how to give and receive feedback (Hersey and Hakel, 1988). In Part 2 we examine the helping relationship itself, highlighting listening and systematic inquiry, reflecting and clarifying, silence, and using confronting behavior. In Part 3 we show how to integrate the techniques into the problem-solving process. In Part 4 we describe the major strategies for helping clients make positive changes in their behavior. We conclude with a new chapter on continuing professional growth.

Changes in the Second Edition

We have expanded to three chapters our introduction to the helping process and the TFA model (Chapters 1–3), and increased our coverage of such crucial topics as the understanding of self and others, awareness of the setting and cross-cultural aspects of helping, and evaluation of the helping process. We have updated the references and recommended readings, and clarified and broadened the role plays and examples. We use the change cycle to integrate the relationship building chapters (Part 2) with the problem solving chapters (Part 3) and the chapters on strategy (Part 4). Some material has been omitted or condensed; for example, Chapter 22 now combines several strategies. The chapters on consultation were dropped as a result of feedback about the first edition.

Special Features

- Examples of using techniques in a variety of settings
- A list of Key Points at the beginning of each chapter
- Role-playing suggestions to help students master chapter content
- Focus checklists of chapter content, to help students critique interviews and get feedback
- Summary checklists that help students integrate all techniques and strategies cumulatively
- References and Recommended Reading at the end of each chapter

Using This Book

Experience has taught us five major ways in which *Helping Relationships and Strategies* can be used to great advantage.

1. *Use all four parts* of the text to gain a basic understanding of the major components of helping relationships and strategies.
2. *Use Parts 1, 2, and 3* to focus primarily on the helping relationship.
3. *Use Part 4* with students who already know many interpersonal helping skills. This section provides an excellent start toward implementing strategies to help clients change behavior.
4. *Use the personal change project* as a basis for conducting the entire course. Students initiate the personal behavior change projects, and course content stems from this experiential base to include more didactic aspects of the helping relationship.
5. Use this text as an excellent overview and refresher for more advanced students before they begin their internship and practicum experience.

Acknowledgments

David would like to acknowledge contributions to the TFA model and its early assessment by Dennis Kinkle, and to the continuing development of TFA Systems™ by Dan Vogler, the world's greatest hobby counselor.

We both appreciate the efforts of the following reviewers: Randy Alle-Corliss, California State University at Fullerton; Nathan Avani, The City University of New York; Diane Coursol, Mankato State University; Paula Dupuy, University of Toledo; James Herbert, Penn State University; Wallace Kahn, West Chester University; and Ralph Mueller, University of Toledo.

David E. Hutchins
Claire G. Cole

References

Frank, J. D. (1981). Therapeutic components shared by all psychotherapies. In J. H. Harvey & M. M. Parks (Eds.), *Psychotherapy research and behavior change*. Washington, D.C.: American Psychological Association.

Hersey, P. W., & Hakel, M. D. (1988). *Leader 1 2 3: A developmental program for instructional leaders*. Reston, VA: National Association of Secondary School Principals.

Hutchins, D. E. (1984). Improving the counseling relationship. *Personnel and Guidance Journal*, *62*, 10, 572–575.

Hutchins, D. E., & Vogler, D. E. (1988). *TFA systems*. Blacksburg, VA: Author.

Norcross, J. C. (1986). *Handbook of eclectic psychotherapy*. New York: Brunner/Mazel.

Paul, G. (1967). Strategy of outcome research in psychotherapy. *Journal of Consulting Psychology*, *31*, 109–118.

Smith, D. S. (1982). Trends in counseling and psychotherapy. *American Psychologist*, *37*, 802–809.

Ward, D. E. (1983). The trend toward eclecticism and the development of comprehensive models to guide counseling and psychotherapy. *Personnel and Guidance Journal*, *67*, 154–157.

CONTENTS

HELPING RELATIONSHIPS

STRATEGIES

Second Edition

Getting Started

In PART ONE, we outline key aspects of this book that will help readers learn about the helping process.

ONE

1 Understanding the Helping Process in This Book

This book will help you learn basic procedures to assist another person in resolving concerns and living more satisfactorily. Understanding the framework of the helping process and commonly used terms used will enable you to use this book more effectively. The *helper* is a person who is trained to assist clients in resolving problems and working through personal concerns. *Clients* are relatively normal people who seek the helper's assistance in resolving problems. The *helping process,* usually occurring over a period of several interviews, is a one-on-one relationship between a helper and client (or group). Minor concerns may be resolved in one to three sessions, whereas more complex problems may extend the involvement for several months.

The helping process differs from typical interpersonal relationships in several ways. In the helping process,

1. The client is primarily responsible for content of the interview.
2. The client receives the helper's total attention rather than engaging in a typical two-way discussion. The helper focuses on the client's concerns instead of expressing personal opinions and giving advice.
3. Client concerns are confidential to the extent allowed by law, professional codes, or contractual agreements.
4. The goal is to assist the client in taking steps to work through concerns and achieve realistic goals—that is, the helping process is goal-oriented.
5. The client can try out new ways of behaving in the secure atmosphere of the helping interview.
6. The client becomes more independent and learns a problem-solving method, which can be applied to a wide variety of situations.

To help you learn the helping process, four major structures organize this book. These include (a) understanding yourself and others (Part One), (b) building blocks in the relationship (Part Two), (c) the problem-solving process (Part Three), and (d) the change cycle. All are interrelated and used in a variety of combinations in Part Four (helping strategies). Major components and how each interacts with others are briefly outlined next and illustrated in Figure 1.1.

Understanding yourself and others

To be an effective helper, you must first understand major aspects of your own behavior and how you interact with others. "Know thyself" is a critical admonition that is as important today as it was in the time of Plato and Socrates. The TFA model and a systematic method of feedback are two critical tools that will be used throughout the book.

TFA triangle

TFA stands for the thinking (cognitive), feeling (affective), and acting (behavioral) components of behavior. *Thinking* approaches are associated with the

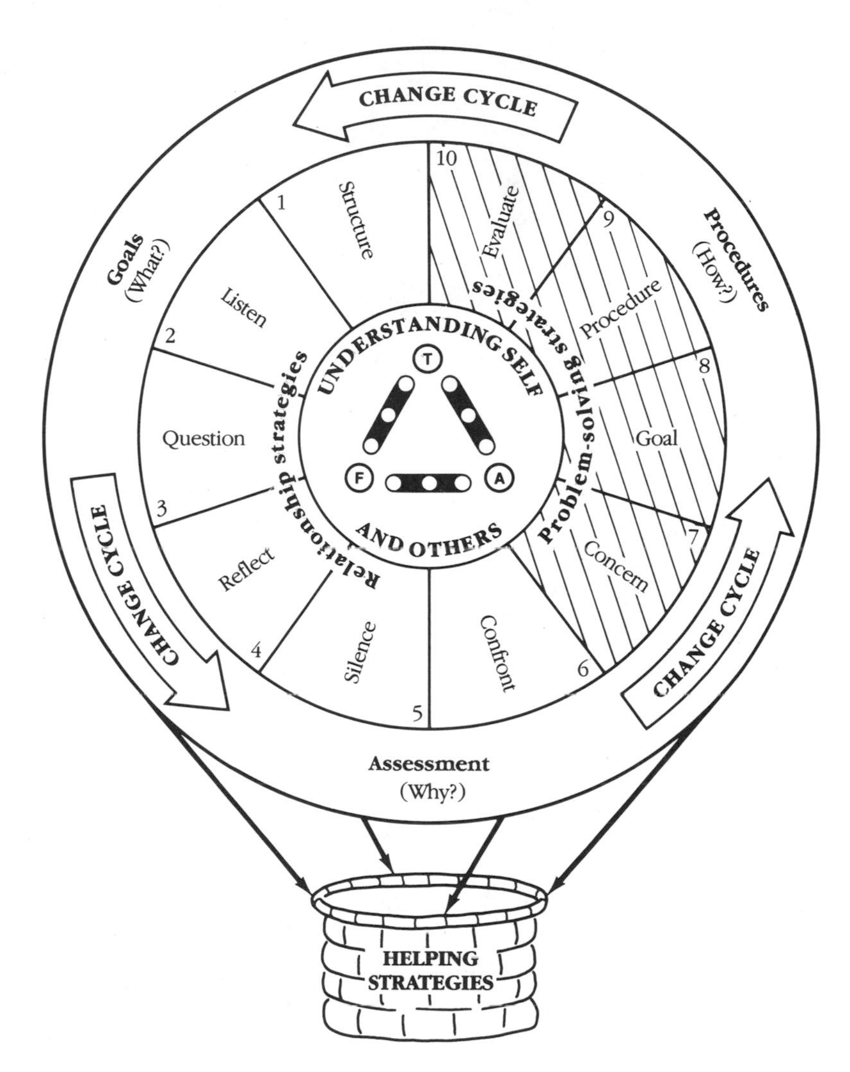

FIGURE 1.1 Four major structures: (a) understanding yourself and others (center), (b) building blocks in the relationship (1–6), and (c) the problem-solving process (7–10). All are interactive in (d) the change cycle and support various helping strategies.

work of Ellis, Beck, Burns, Meichenbaum, Maultsby, and others associated with rational or cognitive theories. *Feeling* approaches are associated with the work of Rogers, Maslow, Perls, and a host of others espousing affective, humanistic, and existential theories. *Acting* approaches are most closely allied with the names of Skinner, Wolpe, Lazarus, Bandura, Krumboltz, Thoresen, and others who use behavioral theories. Many helping techniques and strategies are associated with each of these approaches and reflect the theoretical underpinnings of cognitive, affective, and behavioral orientations (see the recommended readings at the end of this chapter).

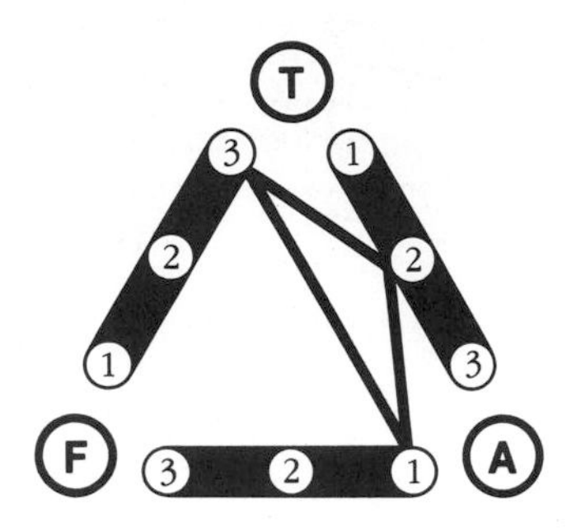

FIGURE 1.2 The TFA triangle with an acting–thinking triad.

Behavior is defined as the interaction among thinking, feeling, and acting (Hutchins, 1984). Since the introduction of the TFA model (Hutchins, 1979), substantial work has been done in research and use of the model in clinical practice. The TFA model has proven to be an extremely useful way of conceptualizing a client's behavior in a specific problem situation. Furthermore, as seen in later chapters, there are major implications for how the helper uses the TFA triangle in the behavior-change process. The TFA model has dual appeal: It is simple enough that beginning helpers and their clients grasp basic components immediately, but it is sophisticated enough that advanced clinical practitioners can use it as a framework for extremely complex problems.

The *TFA triangle* is an open-ended triangular-shaped figure with T (thinking), F (feeling), or A (acting) at each of the vertices. Each side of the TFA triangle is a bipolar scale on which is recorded a range of behavior (Figure 1.2). On the left side of the triangle, behavior ranges from 3 (thinking) to 1 (feeling), with 2 (the midpoint) indicating a blending of these two dimensions. Similarly, on the bottom the range is from 3 (feeling) to 1 (acting). On the right side of the triangle, the range is from 3 (acting) to 1 (thinking).

One's behavior in a specified situation is plotted on each side of the TFA triangle. These points are connected with each other, forming a closed figure called a TFA triad (Figure 1.2). Operationally, the *TFA triad* is a graphic representation of a person's behavior (interaction of thoughts, feelings, and actions) in a specific situation.

By using the TFA triangle, the helper can accurately describe the client's behavior in a specified situation. The triangle is also useful in looking at the helper's behavior and interaction with the client (Chapter 2), identifying client concerns (Chapter 10), establishing goals (Chapter 11), designing procedures (Chapter 12), and evaluating progress (Chapter 13).

Constructive feedback

As we use it, *feedback* is a process in which accurate, descriptive information is given to a person in a caring manner so that it becomes a way of learning more about self and growing in the process. We make the assumption that if you are using this book you are interested in continuing to work on personal development in ways that will enhance your effectiveness with other people. Throughout the book, there are many opportunities to give and receive feedback, and Chapter 3 will help you to do this more effectively.

Building blocks of the effective interview

Part Two, "Building Blocks of the Effective Interview," consists of a group of core skills that are widely recognized as contributing to an effective *working relationship* with a client:

- Structuring the Helping Interviews (Chapter 4)
- Listening for Understanding (Chapter 5)
- Asking Appropriate Questions and Systematic Inquiry (Chapter 6)
- Reflection and Clarification (Chapter 7)
- The Effective Use of Silence (Chapter 8)
- Confrontation (Chapter 9)

These six chapters contain many different techniques that are essential in developing an effective helping relationship and in understanding aspects of the client's behavior and situation (see Figure 1.1b).

Integrating techniques in the problem-solving process

Knowledge of the helper, client, and basic helping skills assists in building an effective relationship. But this is merely the beginning. A *helping relationship* goes beyond good interpersonal interaction to assist the client in resolving concerns or problems in order to live more effectively. Thus, Part Three, "Integrating Techniques in the Problem-Solving Process," which integrates Parts One and Two into the problem-solving process, focuses on a working relationship to help the client bring concerns or problems to successful resolution. The emphasis is on helping to change the client's behavior so the client will think, feel, and act more productively. Figure 1.1c illustrates the chapters in Part Three:

- Identifying Client Concerns or Problems (Chapter 10)
- Establishing Realistic Goals (Chapter 11)
- Designing Effective Procedures (Chapter 12)
- Evaluating, Terminating, and Following-Up (Chapter 13)

The change cycle

Knowing yourself and the client is important (Figure 1.1a). Using good interpersonal skills helps build a good relationship (Figure 1.1b). Integrating these components into a problem-solving phase assists the client in achieving goals

(Figure 1.1c). Although chapters are sequenced as if they occurred in a continuing neat flow, they often are not as neat and clean as presented in the contents. Therefore, we have included the *change cycle* as a major organizer to help you know where you are at any given point in the helping process (Figure 1.1d). Throughout the helping process, you can ask three questions that help tell you where you are:

1. *Why* is the client having difficulty? (Assessment)
2. *What* does the client want to do about it? (Goals)
3. *How* can change be made? (Procedures)

The following explains how the change cycle works.

1. *Assessment*: Why is the client having difficulty?
   To understand the client's concern, the helper must assess the client's behavior and the situation.
   a. What is the client's behavior in terms of the interaction of thoughts, feelings, and actions?
   b. What is the problem situation? Who is involved? Where and when does the situation occur?
   Until these questions are resolved, you are in the assessment phase of the helping process. Various aspects of basic helping skills will assist you in developing a trusting relationship that promotes personal exploration.
2. *Goals*: What does the client want to do about the problem?
   Some kind of change is implied here. Either the client's behavior needs to change, the situation needs to change, or a combination of behavior and the situation need to change.
   a. What are realistic, achievable goals for the client?
   b. What will the client be doing differently that will resolve the problem?
   One of the initial goals of the helper and client is to use all the interpersonal understanding and relationship skills to clearly identify the client's real concern. Once this is identified, the goal shifts to what the client needs to do to resolve the concern. Thus, goals come into play from start to finish in the helping process.
3. *Procedures*: How can changes be made?
   This question addresses procedures that will be used throughout the helping process. The helper must be alert to questions such as
   a. How can the client build on strengths?
   b. How can the client supplement limitations?
   c. How can change strategies be adapted to assist each client?
   Questions such as these affect the initial helper–client interaction, skills used to build the relationship, and strategies used in helping the client to change behavior in the problem situation.

Together, assessment, goals, and procedures of the change cycle translate into three deceptively simple questions: Where are we now? Where are

TABLE 1.1 Integration of the Change Cycle with Relationship and Problem-Solving Strategies

Chapter	Stages of the Change Cycle		
	Assessment	Goals	Procedures
1 Understanding the Helping Process in This Book	✓	✓	✓
2 Thinking, Feeling, and Acting: Understanding Yourself and Others	✓	✓	✓
3 Giving Feedback to Clients and Receiving Feedback as a Helper	✓	✓	✓
4 Structuring the Helping Interviews	✓	✓	✓
5 Listening for Understanding	✓	✓	✓
6 Asking Appropriate Questions and Systematic Inquiry	✓	✓	✓
7 Reflection and Clarification	✓	✓	✓
8 The Effective Use of Silence	✓	✓	✓
9 Confrontation	✓	✓	✓
10 Identifying Client Concerns or Problems	✓	✓	✓
11 Establishing Realistic Goals	✓	✓	✓
12 Designing Effective Procedures	✓	✓	✓
13 Evaluation, Termination, and Follow-Up	✓	✓	✓

we going? How are we going to get there? These questions serve as a framework and guide for helping relationships and strategies from the initial interview through termination.

Each chapter in this book is designed to address one or more of these central questions. Table 1.1, in which specific chapters are listed as they relate to the three questions in the change cycle, shows the organization of this book.

Guidelines for using this book

In Parts Two and Three, the process of using this book will be as follows:

1. Read the chapter. In each chapter, *focus* on specific content in trying to improve your behavior as a helper.
2. Practice the focus activities. Conduct a number of mini-interviews with a *coached client* who agrees to help *you* improve your behavior in this area. The mini-interviews are very short (three to ten minutes). Your goal is to focus on specific content, practicing it until it comes smoothly. Ultimately, it will come "naturally."
3. Use the "Focus Checklist" that appears at the end of most chapters. These checklists help you assess your performance during your focused practice in the mini-interview. You'll be able to see what you do or do not include, what you do well, and where you need more practice.
4. Repeat steps 1–3 as often as necessary for the focus behavior in each

chapter to occur naturally, without having to think about it. Any new behavior is likely to be somewhat artificial, stiff, or awkward at first. Remember how hard it was to write, type, play sports, or tie your shoes the first few times?

5. Review and integrate previous skills. When you have a new focus activity, first practice it. Then review and integrate all previous skills into the mini-interview. The result is a cumulative process in which you become increasingly more adept at using skills *appropriately* in an interview situation.
6. Complete the "Summary Checklist." Beginning with Chapter 6, each chapter will have a "Summary Checklist." These checklists will help you to appropriately integrate all previous content *after* you have practiced the focus activity. You will not use every skill all the time. The ultimate goal is to use the relevant skill in a manner that lets you conduct an effective helping interview.

Helping strategies

Part Four, which has 10 chapters, addresses numerous strategies that can help the client in the process of changing behavior. With varying combinations of helping skills and techniques, helpers can use hundreds of different strategies to assist clients in the change process. These will need to be adapted to fit each client for a specific situation, but the beginning helper can draw from several useful strategies:

- Running Accounts and Recording Behaviors (Chapter 14)
- Autobiography and Personal History (Chapter 15)
- Structured Information Gathering: Questionnaires, Open-Ended Sentences, and Structured Interviews (Chapter 16)
- Charting Behavior and Recording Progress of Behavior Change (Chapter 17)
- Contracting (Chapter 18)
- Positive Self-Talk (Chapter 19)
- Imitation and Modeling (Chapter 20)
- Role Playing (Chapter 21)
- Using Additional Strategies: (Chapter 22)
- Where Do I Go from Here? Continuing Development for Helping Professionals (Chapter 23)

Each strategy is critiqued in terms of strengths and limitations of working with various clients. Furthermore, guidelines are provided for how to use the strategies with different behavior and where the strategies may be especially useful in different steps of the change cycle.

The final chapter—"Where Do I Go from Here? Continuing Development for Helping Professioinals"—is designed as a springboard to help you continue your personal and professional development, moving on from this book.

A personal change project

A personal change project encourages you to use many of the techniques and strategies in this book as you engage in responsible experimentation in changing aspects of your personal behavior (how you think, feel, and act). The project makes the course come alive when *you* actually use many of the skills in this book on yourself.

Choosing a personal change project

Think about some aspect of your personal behavior that you would really like to change. The project should be important to you, yet one that you can share with others. The project must depend on *changes you can make*. For example, if you are having trouble interacting with another person (e.g., employer, employee, spouse, or child), you would focus on how you can change instead of what the other needs to do. Your project should involve personal behavior that you can decide to change.

Personal change projects have helped individuals to

- Interact more effectively with others in social situations.
- Remember names when introduced to people in a group.
- Improve interactions with children in a family.
- Start and continue conversations with new acquaintances.
- Resolve conflicts between family members or friends.
- Enjoy leisure time and related activities without guilt.
- Become more assertive in positive, responsible ways.
- Improve self-concept.
- Stop smoking.
- Lose weight.
- Eat more nutritionally healthy and balanced meals.
- Start and maintain a regular exercise program.
- Learn to relax without TV, drugs, or other people.
- Learn a new language.
- Learn to use a word processor, computer, or software program.
- Develop and stick to a budget.
- Organize and manage time more effectively (home and job).
- Learn to study or read more effectively and efficiently.
- Eliminate self-defeating attitudes.
- Overcome procrastination.

The impact of these personal change projects is increased if some class time each week is devoted to discussion and analysis. We suggest that instructors not grade your success or failure of the project. This encourages you to engage in responsible experimentation with a variety of techniques and strategies. We do suggest a very brief (half- to one-page) written summary in which you record important events (as described next).

Getting started and keeping track of events

By the second class meeting, submit a brief (one-page) summary of the items outlined below for discussion. This will be your first report for the behavior-change project. Your first report on goals, procedures, and strategies will be rough. It is likely that you will make changes and revisions during the course. Observing what you do and why you do it are important parts of the learning process.

First report

1. What is your major concern for this project in terms of
   a. Your behavior (how you think, feel, and act), and
   b. The situation (who is involved, when and where)?
2. What is your major realistic goal for this personal behavior-change project during the term? The target date should be one or two weeks before the end of the term.
3. What major procedures/strategies do you expect to use in working toward your goal? At this point, you can benefit from skimming over Part Four (strategies), noting activities that could be used in your project.
4. How will you evaluate the outcome of the project?

Weekly behavior-change summary

Beginning the second week and continuing for six to twelve weeks, bring to class a one-page summary of the past week's events related to your personal behavior-change project that includes six items:

1. Primary goal for the term.
2. Major goal(s) for the past week.
3. Procedures used. Indicate what methods you used to try to change your thoughts, feelings, and actions. Be specific.
4. Results during the week. List specific events that affected your progress. Here are some examples:
   a. I succeeded in getting to work on time four out of five days. My motivation was very high. I also was complimented by my boss.
   b. The weekend was a disaster. A friend I hadn't seen in three years came to town, and I did not do any of [the project] from Friday until Monday.
   c. It was our anniversary. We splurged by eating out at an expensive restaurant and completely blew our budget for the month.

d. At first I detested putting notes all over the house reminding me to put things away, but after about five days I found my attitude was changing. Cleaning up the house only took fifteen minutes a day instead of the hours I thought it would.

5. Prescription for next week. Assume you are the *helper* for a client who has the same experiences that you did. What would you recommend this client do during the next week? Why?
6. Implications for the helping process. List implications for the helping process that you learned from your behavior-change experience.

We strongly recommend that you keep a journal during the week in which you describe specific events that occur related to your project. Include specific aspects about your thoughts, feelings, and actions (e.g., enthusiasm, doubt, skepticism, commitment, successes, failures, attitude, and specific events) that promoted or prevented goal achievement. Such notes can help you in writing your weekly behavior-change summary. They can also be useful if you are asked to prepare a one-page summary of your experience during the term.

To process the behavior-change experiences in class, two variations have been used successfully. First is that you get together in small groups to share your experiences. Then perhaps two or three of you from the class may have had especially interesting or unusual events occur that are shared with the entire class. Another option is to devote part of each class and have all or some of you share your projects with the entire class each week.

References and recommended reading

Bandura, A. (1986). *Social foundations of thought and action: A social cognitive theory*. Englewood Cliffs, NJ: Prentice-Hall.

Beck, A. T. (1976). *Cognitive therapy and emotional disorders*. New York: New American Library.

Burns, D. D. (1980). *Feeling good: The new mood therapy*. New York: Morrow.

Corey, G. (1991). *Theory and practice of counseling and psychotherapy* (4th ed.). Pacific Grove, CA: Brooks/Cole. This is an excellent overview of major theories of counseling and psychotherapy, presented in a concise manner.

Ellis, A., and Grieger, R. (1977). *Handbook of rational-emotive therapy*. New York: Springer.

Grieger, R., & Boyd, J. (1980). *Rational-emotive therapy: A skills-based approach*. NY: Van Nostrand Reinhold.

Hutchins, D. E. (1979). Systematic counseling: The T–F–A model for counselor intervention. *Personnel and Guidance Journal*, *57*, 529–531.

———. (1984). Improving the counseling relationship. *Personnel and Guidance Journal*, *62*, 572–575.

Krumboltz, J. D. (1980). A second look at the revolution in counseling. *Personnel and Guidance Journal*, *58*, 463–466.

Lazarus, A. A. (1981). *The practice of multimodal therapy*. New York: McGraw-Hill.

Maslow, A. (1986). *Toward a psychology of being* (rev. ed.). New York: Van Nostrand Reinhold.

Maultsby, M. C. (1984). *Rational behavior therapy*. Englewood Cliffs, NJ: Prentice-Hall.

Meichenbaum, D. (1977). *Cognitive behavior modification: An integrative approach*. New York: Plenum.

———. (1986). Cognitive behavior modification. In F. H. Kanfer & A. P. Goldstein (Eds.), *Helping people change: A textbook of methods* (3rd ed.) (pp. 346–380). New York: Pergamon Press.

Perls, F. (1969). *Gestalt therapy verbatim*. Moab, UT: Real People Press.

Rogers, C. (1961). *On becoming a person*. Boston: Houghton Mifflin.

Skinner, B. F. (1979). *The shaping of a behaviorist*. New York: Knopf.

Thoresen, C. E. (Ed.). (1980). *The behavior therapist*. Pacific Grove, CA: Brooks/Cole.

Wolpe, J. (1969). *The practice of behavior therapy*. New York: Pergamon Press.

Thinking, Feeling, and Acting: Understanding Yourself and Others

2

Key points

1. Knowledge of self is essential for working effectively with others.
2. Behavior is the interaction of thoughts (T), feelings (F), and actions (A).
3. Behavior is dimensional, not static. It may change in different situations.
4. Assessing behavior in a specific situation provides useful clues to understanding self and working with others in a helping relationship.

The thinking–feeling–acting (TFA) approach

The TFA (thinking, feeling, acting) system (Hutchins, 1979, 1982, 1984) is a unique and practical method of integrating the three major approaches in the helping profession. The cognitive or rational (*thinking*) approach has been influenced by the work of Ellis (rational-emotive therapy), Beck (cognitive therapy), Maultsby (rational behavior therapy), Meichenbaum (cognitive modification), and others. The affective or humanistic (*feeling*) approach has been influenced by the work of Rogers (client-centered or person-centered therapy), Perls (Gestalt therapy), Maslow, and a host of phenomenological, humanistic, and existential writers. The behavioral (*acting*) approach has been influenced by the work of Skinner, Bandura (behavior modification), Wolpe (behavior therapy), Lazarus (behavior therapy and beyond), Krumboltz and Thoresen (behavioral counseling), and others. Individually, these are different ways of understanding, explaining, and changing human behavior. Each has made tremendous contributions to understanding how people function. However, the greatest single current trend in the helping profession is toward integration (Smith, 1982; Ward, 1983) of various approaches.

Debates continue about the role of thoughts and feelings and which are primary (Lazarus, 1984; Zajonc, 1984). Hutchins (1984) believes that thoughts, feelings, and actions are all part of human behavior and each is crucial to understanding and working effectively with people. In any given situation, thoughts, feelings, or actions may become a trigger for different kinds of behavior. Furthermore, no single dimension of behavior (thoughts, feelings, or actions) is pure or discrete. Rather, thoughts, feelings, and action dimensions embody *some* components of the others. Ellis supports this notion in responding to the TFA model:

> It is doubtful whether any of the "separate" terms, *emotion, cognition, and behavior,* can be defined in their own exclusive right, since each contains salient and essential elements of the other two. When theories of counseling and psychotherapy are divided into . . . Thinking–Feeling–Acting categories, most major schools can be fairly accurately placed in one of these three categories. (1982, p.7)

COMPREHENSIVE ABC MODEL		
A **Antecedents** (Situation or context)	**B** **Behavior** (Thoughts, feelings, and actions)	**C** **Consequences** (Payoffs and tradeoffs)
Under what circumstances does the behavior occur? 1. Where? 2. When? 3. With whom?	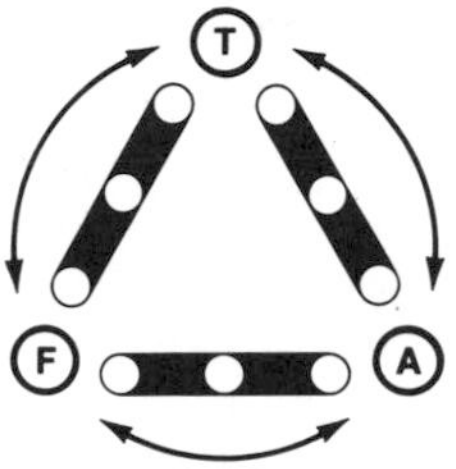 T F A	1. Payoffs/ strengths 2. Tradeoffs/ negatives a. Immediate b. Short-term c. Long-term (of 1 and 2)

FIGURE 2.1 Comprehensive ABC model.

The comprehensive ABC model

The TFA model, which follows this section, is best understood in terms of a comprehensive ABC model.[1] As shown in Figure 2.1, *A* stands for antecedents, *B* for behavior (thoughts, feelings, and actions), and *C* for consequences.

A is the antecedent event, external situation, or context in which one's behavior occurs. It is critical to examine the situation because nothing occurs in isolation or has meaning outside of some framework or situation. As an example, think of *your* behavior while reading this text, in a crisis situation, relaxing, socializing with a good friend, or in a particular job. The situation is examined through exploring specific aspects of where, when, and with whom one's behavior takes place.

In each of these situations, one's behavior (*B*) may change in important ways. For some people, behavior (the interaction of thoughts, feelings, and actions) may vary dramatically in different situations, whereas for others there may be remarkable consistency from one situation to another.

Consequences (*C*) are the results of one's behavior in a specified context. As a result of one's thoughts, feelings, and actions in a particular situation, a variety of consequences will occur: Some are positive (payoffs), whereas others are more negative (tradeoffs); some are immediate or short-term, whereas others are of much longer duration or even permanent. One's behavior cannot be fully understood without examining the interplay between the antecedents—interaction of thoughts, feelings, and actions—and the consequences of behavior. The remainder of this chapter examines behavior in

[1] Thanks to Wally Kahn for his input on the ABC model. Also note that Albert Ellis (1982) uses ABC somewhat differently from this.

some detail. Chapters 10 (concerns), 11 (goals), and 12 (procedures) explore other aspects of the comprehensive ABC model.

Assessing thinking, feeling, and acting components of behavior

One's thoughts, feelings, and actions can be assessed in two major ways. In this book, we will use the TFA triangle as outlined next. A second method is to use the Hutchins Behavior Inventory (HBI) (Hutchins, 1984; Hutchins and Mueller, in press; Mueller, Hutchins and Vogler, 1990).

The *TFA triangle* is a three-sided figure open at each of the three vertices, which represent thoughts (T), feelings (F), and actions (A). Each side of the triangle is a bipolar scale. On the left side of the TFA triangle is a thinking–feeling (T–F) scale; on the bottom is a feeling–acting (F–A) scale; and on the right side is an acting–thinking (A–T) scale (Figure 2.2).

You can assess your own behavior (and that of others) by using the TFA triangle. Try it now on yourself. Think about your behavior in the role of a helper and ask the three critical questions that follow. Record your answer on the blank TFA triangle (Figure 2.2b). (The completed TFA triangle of Figure 2.2a is used as an example.)

1. T–F (left side of triangle): Is your behavior based more on thinking (3), feeling (1), or is it about in the middle (2)? Mark an *X* in 1, 2, or 3 on the left side of the TFA triangle.
2. F–A (bottom of triangle): In the same situation, is your behavior based more on feeling (3), action (1), or is it about in the middle (2)? Mark an *X* in 1, 2, or 3 at the bottom of the TFA triangle.
3. A–T (right side of triangle): In the same situation, is your behavior based more on action (3), thinking (1), or is it about in the middle (2)? Mark an *X* in 1, 2, or 3 on the right side of the TFA triangle.
4. Now, connect the points you marked with an *X,* to form a closed triangle. This closed inner triangle is called the *TFA triad* (Figure 2.2a).

As you can see, the basic process is easy. The next step gives you much more information about yourself (as well as others with whom you interact).

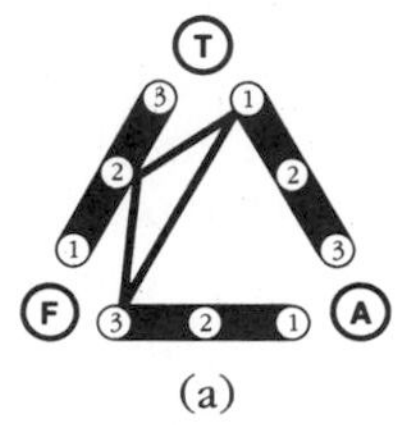

(a)

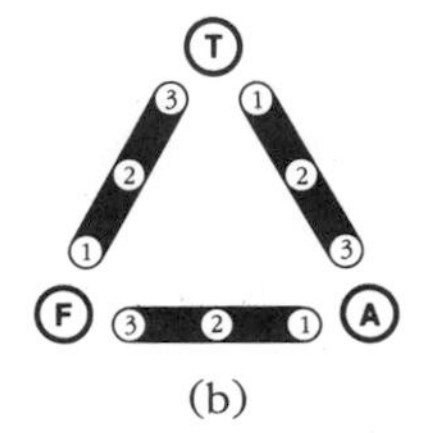

(b)

FIGURE 2.2 (a) A completed thinking–feeling triad. (b) Complete this triad for yourself in the role of a helper.

I'm logical and rational.
I balance my thoughts with my feelings.
I'm very caring.
I'm sensitive to thoughts of others.
I get overinvolved with others' feelings.

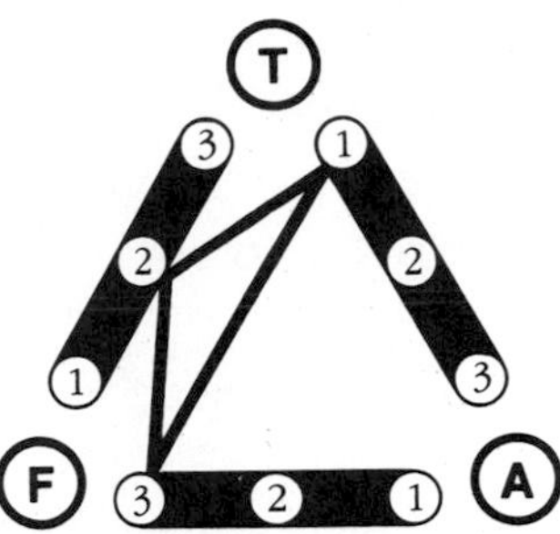

I plan ahead.
I think about alternatives.
I try to anticipate answers to questions.
I'm hesitant to do things until I know for sure what to do.

I'm sensitive to others' feelings.
I care a lot about others.
I can empathize with people.
I'm not very assertive.

FIGURE 2.3 The TFA triangle with the individual's triad plotted. On each side of the triangle, notes specify *why* the particular point on each side of the triangle was selected as being characteristic of behavior in the situation.

Describing specific behavior

After first completing the TFA triad, go back over the questions *for each side* of the triangle and ask the key questions:

> What are specific aspects of my behavior that led me to answer as I did?
> What are the reasons I answered (1, 2, or 3) as I did?

Write down these specifics. To illustrate the process, let's assume that your TFA triad is like the one in Figure 2.3. This person answered the key questions as shown below and made *notes,* which summarized personal behavior in the situation at various points on each side of the triangle. When using the TFA triangle, it is helpful to jot down specifics as in Figure 2.3. In later chapters (10–13), it will be clear why this is done.

On the T–F (left side), you marked that you are about in the middle (2) between thinking and feeling. Your list of specific details might look like this:

> I balance my thoughts with my feelings.
> I'm very caring.
> I'm sensitive to the thoughts of others.
> I get overinvolved with others' feelings.
> I'm logical and rational.

On the F–A (bottom side), you marked (3)—indicating you are mostly feeling. Specific aspects about your behavior are as follows:

> I'm sensitive to others' feelings.
> I care a lot about others.
> I can empathize with people.
> I'm not very assertive.

On the A–T (right side), you marked (1)—indicating you are mostly thinking. Specific aspects include these:

I plan ahead.
I think about alternatives.
I try to anticipate answers to questions.
I'm hesitant to do things until I know for sure what to do.

Note that some of these specifics mentioned for each side of the TFA triangle may be associated with both positive and negative consequences (*payoffs* and *tradeoffs*). These aspects will be explored more fully in Chapters 10–12.

Interpreting the triad pattern

When you have completed the TFA triad and written several specific, descriptive aspects about your behavior, it will likely be different from the one in Figure 2.2a. Now determine what the TFA triad means (i.e., what the TFA triad means about your behavior as a helper). With a partner or small group, respond to items 1–4. Examples of the kinds of things one might say follow each question. These examples are for a person with the TFA triad shown in Figure 2.2a.

1. *Describe what the TFA triad means about your behavior in this specific situation.* Give a general overview here: "This TFA triad means that I think a lot about what I'm going to do. I like to be pretty sure of myself. I also care a lot about others. I'm sensitive to their feelings. But I guess I don't take much action. I'm a thinking person and feel very deeply about others."
2. *What are the payoffs (strengths, or positive aspects) of your behavior pattern in this situation?* "I'm caring and empathic. Others feel comfortable confiding in me. I think things through well and don't jump into things without being aware of the consequences."
3. *What are the tradeoffs (limitations, or negative aspects) of your behavior pattern in this situation?* "Well, sometimes I don't get around to doing things as much as I could. I'm not very assertive. I'm usually the last one to speak up. I don't always follow through and complete tasks as well as I would like."
4. *What are the implications of my TFA triad for the helping relationship?* "I think I could work to become more assertive in getting things done. I do come across as genuinely interested and sincere, but I may need to be more active. I wonder if I work better with some people than others. I might have difficulty working with people who are much more action-oriented than I. I think I need to be aware of different people and situations and how I work with them."

Activity

Practice this TFA exercise with different people. Using friends, classmates, or coached clients may be best at first. Vary the specific situation used. Have your partners describe their behavior in a variety of situations: for instance, where they care deeply about another person, in the heat of anger, during a crisis, at a party, as a student, interacting with an employer or employee, being refused a job, or at a social service agency. *It is important to have the person specify the specific situation* and then respond to questions in terms of this situation. You'll find that this is a powerful method of obtaining information about others that is extremely useful later on in the interview sequence (see Chapters 10–12 on concerns, goals, and procedures).

Questions to ask following interviews using the TFA triangle

1. What were your reactions to the TFA process?
2. How did the use of the TFA process help you find out more information about the client?
3. How were the client's thoughts, feelings, and actions consistent or inconsistent with each other?
4. Looking at the written responses at each side of the TFA triangle, what are the implications for the helping relationship?
5. As a helper, what kinds of behavior do I need to be aware of in working with a person in this situation?
6. What kinds of strengths and limitations do I have?
7. How do these relate to my being able to work with certain clients or situations?

Use of the TFA model is a new and powerful approach to the helping process. Key aspects of the TFA model are briefly as follows:

1. Behavior is the interaction of thoughts, feelings, and actions in a specified situation. Thoughts, feelings, and actions are dealt with as a gestalt and *how they interact with each other* rather than as separate entities.

2. Thoughts, feelings, and actions are not usually discrete. They occur in relationship to each other with varying degrees of overlap and interaction. The TFA triangle lets us examine this interaction in almost any situation.

3. Behavior is dimensional or situational. While a person may have typical or characteristic ways of behaving, specific situations may result in radically different behavior patterns (Clow, Hutchins and Vogler, 1990). The helper needs to be aware of differences in the client's behavior in specific situations in order to help design effective behavior-change strategies. The comprehensive ABC model provides a context in which thoughts, feelings, and actions of the client can be explored. Figure 2.4 on page 22 shows how one person may behave differently in four specific situations: at work, as a group leader, when confronted by a supervisor, and when upset with a child.

4. A stimulus in any thinking, feeling, or acting dimension of behavior may trigger reactions in other dimensions. Thoughts may trigger feelings as well as actions; feelings may trigger thoughts as well as actions; and actions may trigger thoughts as well as feelings (Figure 2.5 on page 22): (a) (thought) I must have done poorly on the exam leads to *feelings* of frustration and more intense study. (b) Feeling sad about the loss of a friend may lead one to think about one's own mortality and to do some things that had been previously

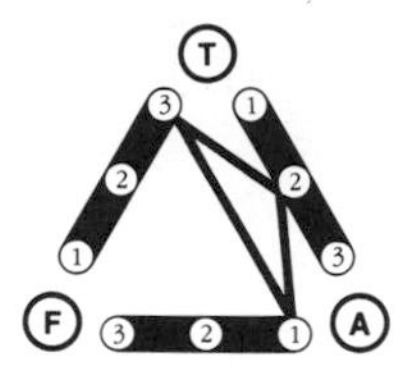

(1) Triad at work
(thinking–acting)

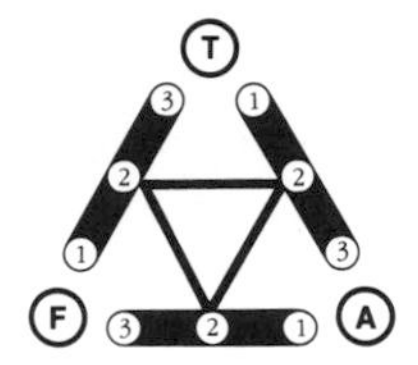

(2) Triad as a group leader
(midpoint triad)

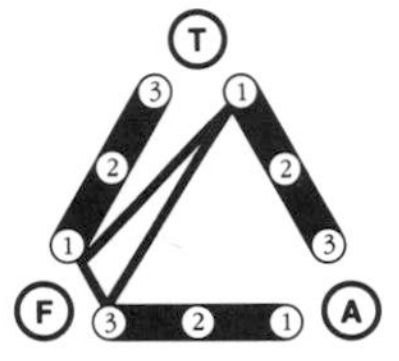

(3) Triad when my supervisor confronted me about my poor work performance (feeling–thinking)

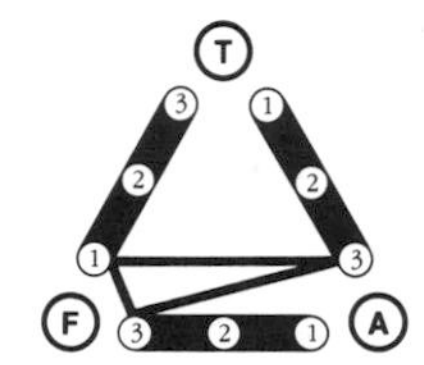

(4) Triad when I was upset with my child's failure to start homework (feeling–acting)

FIGURE 2.4 Four TFA triads illustrating the same client's behavior pattern in four situations.

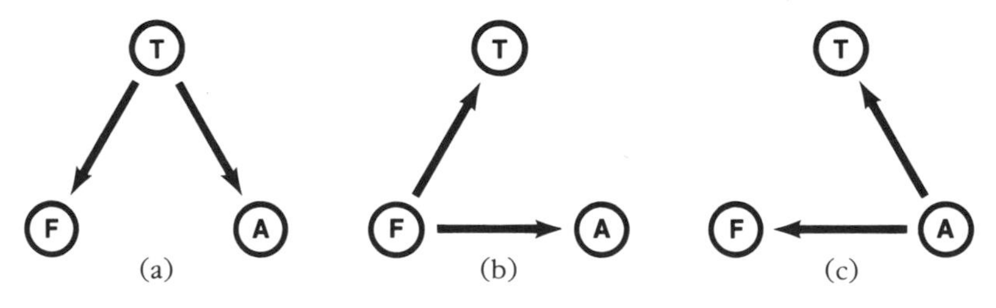

FIGURE 2.5 Various kinds of stimuli may trigger different kinds of responses. (a) Thoughts can trigger feelings and actions, (b) feelings can trigger thoughts and actions, and (c) actions can trigger both thoughts and feelings.

ignored. (c) One's recent promotion (action) may lead to feelings of exhilaration and positive thoughts about options that can be taken. This is especially important in understanding behavior and in designing behavior-change strategies.

Four major TFA patterns

Twenty-seven different triads can be formed in the TFA triangle because of the three points on each side ($3 \times 3 \times 3 = 27$). As you engage in responsible experi-

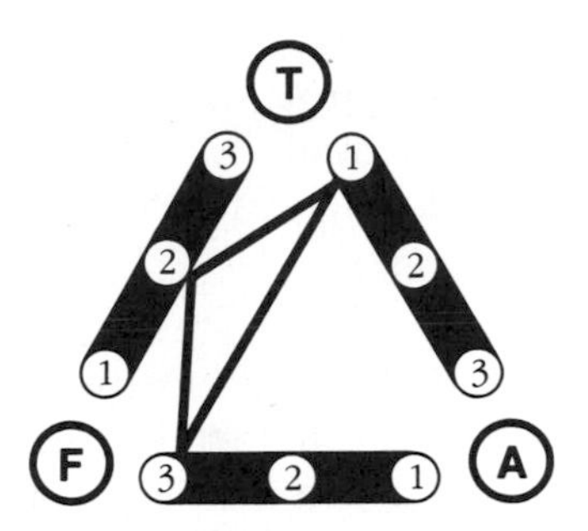

FIGURE 2.6 A thinking–feeling (T–F) triad.

mentation with the TFA triangle, you'll notice that there are patterns with consistent themes.

The twenty-seven triads can be grouped into four major triad groups. Each has variations that relate to dominance or blending of the thinking, feeling, and acting dimensions. As an example, on the left (T–F) side of the TFA triangle, points 3 and 1 reflect primarily (3) thinking or (1) feeling, while 2 may indicate a relative blending or ambivalence of thoughts *and* feelings. The other sides are interpreted in the same manner. The four major TFA groups and their characteristics are briefly outlined next. These descriptions can serve as general guidelines for interpreting the four major triad patterns, especially when coupled with information obtained from using the TFA triangle in interviews.

Note that when comparing the individual's TFA triad with the four major triad groups, one may have a triad that varies by one or two points from the major groups. When this happens, one can expect behavior that is more representative of these points. For example, with the first (T–F) triad, if the midpoint on the left were moved toward T, one would expect there to be more thinking (Figure 2.6). Similarly, if the midpoint were moved toward F, there would be more feelings and emotions. Finally, if either the T or F points were moved toward A, one would expect the person to be more action-oriented than the description indicates.

T–F: Thinking–feeling (left side of triangle)

The thinking–feeling (T–F) pattern is one in which persons think a lot and are stimulated by ideas, what could be done, consequences of events, and so forth. They are frequently insightful, curious, and clever. They are concerned about what other people think. They may have strong feelings about the situation. Strong interaction between thoughts and feelings is very likely. Thoughts influence why people feel as they do, and these feelings, in turn, affect the thinking process. Frequently, there is a continuing, cyclic, positive, and/or negative interaction between thoughts and feelings or relative ambivalence between the two.

Payoffs, or strengths

Both thoughts and feelings have a great impact on behavior. Persons with this pattern can be extremely thoughtful, logical, caring, sensitive, and concerned, and may become passionately involved in issues.

Tradeoffs, or limitations

Persons with this pattern may become involved in a number of projects undertaken in the enthusiasm of the moment. Frustration occurs as deadlines approach and when ideas and feelings conflict. A lack of any triad pattern in the acting area suggests that action to complete tasks is often not matched by the initial feelings of excitement about ideas.

F–A: Feeling–acting (bottom of triangle)

In this group, behavior is feeling and action-oriented (Figure 2.7). Persons with this pattern are likely to react emotionally to events and take what they believe is appropriate action. Strong feelings may trigger actions, or actions may trigger intense feelings. Such patterns may result in compassionate, warm, appreciative, sincere, and spontaneous behavior. The situation at the moment is likely to determine whether one's behavior is positive or negative. Feelings and actions can be integrated with each other or ambivalent.

Payoffs, or strengths

Persons with this pattern tend to be open and affectionate. They trust others and enjoy social interaction. At times their impulsive action may be taken at great personal risk, leading to unselfish and even heroic behavior.

Tradeoffs, or limitations

Persons with this pattern tend to act without considering all the consequences of their behavior, as suggested by the absence of the triad in the thinking area of the TFA triangle. As a result, much time may be spent in making up for or redoing things that occurred because of impulsive actions.

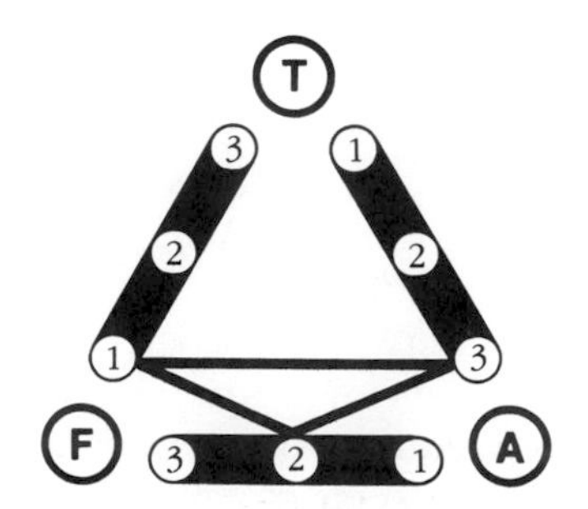

FIGURE 2.7 A feeling–acting (F–A) triad.

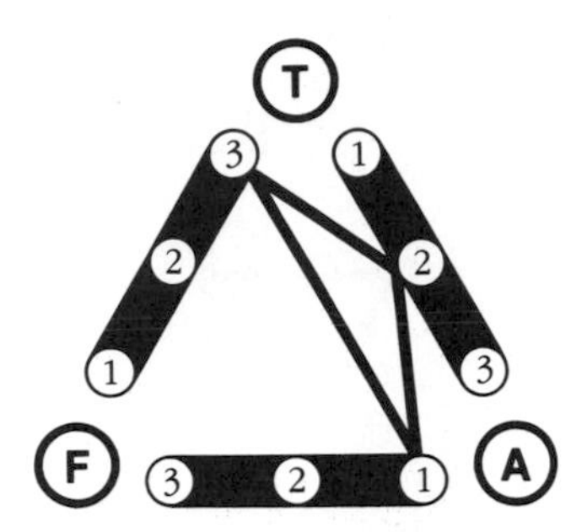

FIGURE 2.8 An acting–thinking (A–T) triad.

A–T: Acting–thinking (right side of triangle)

This group is characteristic of persons who carry their ideas successfully through to completion (Figure 2.8). They are often insightful, discerning, assertive, persistent, and persuasive. They see things that need to be done, develop a plan of action, and act decisively to finish tasks. Behavior can be integrated between actions and thoughts, or it can be ambivalent.

Payoffs, or strengths

Persons with this pattern tend to think and act decisively. They plan and organize activities, consider options, and take action needed to complete tasks on time. They are known as "can do" people who get things done.

Tradeoffs, or limitations

Persons with this pattern may be impatient with others who hesitate. They may appear insensitive and aloof, rolling over others to get the job done. As the absence of the triad suggests, they may fail to acknowledge the importance of others' feelings in making decisions.

T–F–A: Thinking–feeling–acting (midpoint on the triangle)

When midpoints on each side of the triangle are connected (Figure 2.9 on page 26), there is a balance of thoughts with feelings and actions. Such people tend to be thoughtful, understanding, and cooperative, yet purposeful and task-oriented. This is often the *best* problem-solving pattern in a helping relationship because all aspects of a situation are considered before taking action.

Payoffs, or strengths

Persons with this pattern tend to look at things from many different points of view, relate well to a wide range of feelings, and act in the best interest of all. They act in democratic ways and make good facilitators, mediators, and

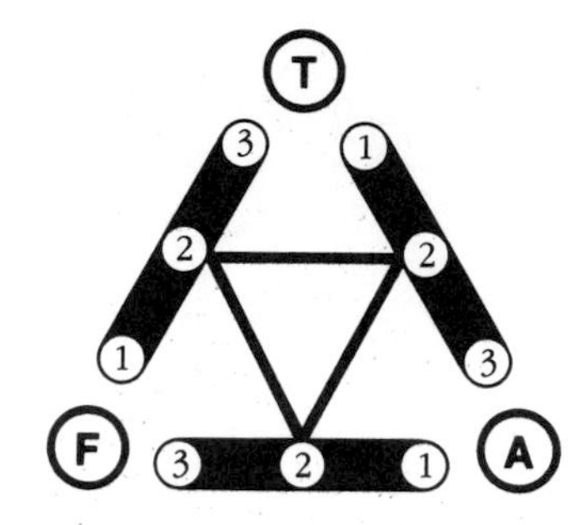

FIGURE 2.9 A midpoint (T–F–A) triad.

arbitrators because of their ability to see and understand all dimensions of the situation. They tend to operate best in longer-term situations, which allow adequate consideration of all variables.

Tradeoffs, or limitations

Persons with this pattern often tend to compromise efficiency in what they do. They think things through thoroughly while considering feelings and emotions of others and what the impact of potential actions may be. They appear, at times, to be delaying or avoiding the situation. In the worst case, they may become stuck, dead center, when there is conflict between thoughts, feelings, and actions. This is especially true if some people might get hurt; they may delay making hard decisions until the balance changes, when it may be too late for effective action. They function best in long-term situations, which permit adequate consideration of all dimensions of behavior.

The T–F–A midpoint triad and problem solving

An immediate (not necessarily the ultimate) goal of the helping process is to help the client reach a midpoint (balanced) triad on the TFA triangle (Figure 2.10). This midpoint triad suggests that the client can cognitively "see" all points of view and options, understand how feelings and emotions come into play, and take appropriate action.

Variations of the major patterns

Two groups of triads vary from the groups already mentioned.

One-dimension midpoint triad

The first variation is when two points on the triad are *both* at one of the vertices (two points at T, F, or A; see Figure 2.10). When this occurs, the T, F, or A dimension is likely to have a very great impact on how the person functions. The third (less dominant) point is at a midpoint position opposite the dominant T, F, or A vertex.

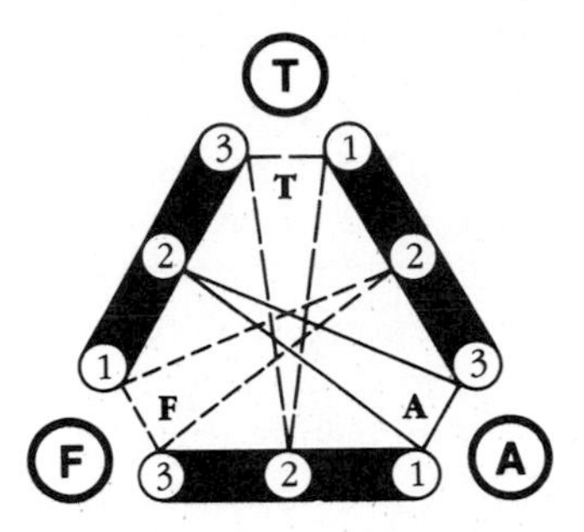

FIGURE 2.10 Three triads with an emphasis on thoughts, feelings, or actions, coupled with a midpoint on one side of the triangle.

Payoffs

These relate to the areas where someone has two points at a T, F, or A vertex. The person emphasizes dominant thoughts, feelings, or actions in the triad while integrating some aspects of the midpoint position.

Tradeoffs

The person may emphasize dominant dimensions of T, F, or A to such an extent that other, midpoint, aspects are not fully explored.

Compartmental triad

The second variation is where *all three points* are at the T and F, and A positions (Figure 2.11). With this triad pattern, one's behavior is called *compartmental*. The person may function very well with this pattern as long as there is an overall plan. However, when the plan is upset or interrupted, behavior is likely to be compartmentalized. Thoughts, feelings, and actions occur as if they were separated and unrelated to other dimensions. Behavior in each of these areas may appear intense and out of context. The sequence of T, F, and A may be in any order, and behavior is likely to appear inappropriate until the person initiates another plan in which thoughts, feelings, and actions work together.

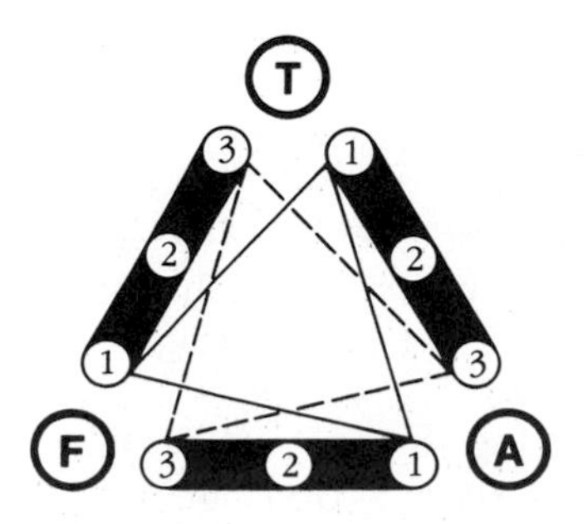

FIGURE 2.11 Two compartmental triads emphasize thoughts, feeling, and actions at different times.

Payoffs and tradeoffs of the compartmental triad

Payoffs appear when a person has focused strongly on all aspects of each thinking, feeling, and acting dimension. As a result, these dimensions are reflected in an integrated pattern of behavior that shows attention to most of the important details of the situation.

Tradeoffs occur when the person's initial plan is interrupted or has not been developed fully. Thoughts, feelings, and actions occur as compartmentalized or isolated pieces of behavior that are out of proportion to the situation.

Payoffs and tradeoffs for each client

Although each of the major triad groups tends to have general payoffs and tradeoffs, it is most helpful to probe into each client's behavior to determine what these are in the problem situation. Each individual has characteristic behavior patterns that tend to be used in specified situations. If one's behavior can be identified through work with the helper, both strengths and limitations for the client can be used in the process of changing behavior, as illustrated in Chapters 10–12.

Summary

1. Behavior is the interaction of thoughts, feelings, and actions.
2. Behavior is dimensional. Because it varies as a function of different circumstances, one's behavior is best understood in terms of the specific situation.
3. An event that occurs primarily in a thinking, feeling, or acting dimension usually triggers reactions in other areas.
4. Knowledge of self is critical to working effectively with others in a helping relationship.
5. In any given situation, one can use the TFA triangle to assess thinking, feeling, and acting components of behavior.
6. The TFA assessment leads to analyzing payoffs/strengths, tradeoffs/limitations, and implications for working effectively with others.

Role play

1. Recuit a coached client to assist you in learning and practicing your skills.
   a. Have your coached client assume some kind of problem situation (e.g., difficulty in a relationship with a friend, spouse, child, employer, employee, or colleague at work) for which help is sought.
   b. After identifying the specific problem, go through the TFA triangle with the client as described in "Assessing Thinking, Acting, and Feeling Components of Behavior."

(1) Plot the TFA triad.
(2) Have the coached client describe the specific behavior.
(3) Work with your coached client to determine the meaning of the triad.
(4) Discuss payoffs, tradeoffs, and implications of this pattern of behavior in the specific (problem) situation.

2. Complete a TFA triangle for *you* as the helper working with the client in step 1. Compare the TFA triad of the client and yourself. Then, with a friend taking the course, discuss the strengths and limitations of these two triads and potential implications for
   a. The initial client–helper relationship.
   b. Similarities and differences in TFA orientations.
   c. How the helper may need to adapt in order to join with the client in understanding the client's perspective.

Make copies of the TFA triangles (Figure 2.12) and use them as you interact with others in aspects of the helping profession.

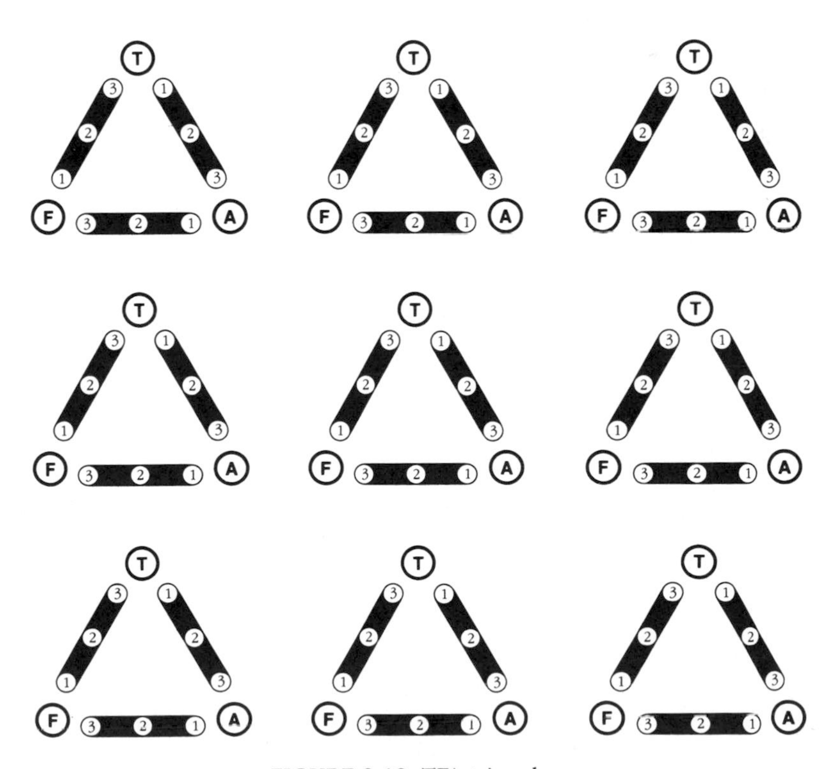

FIGURE 2.12 TFA triangles.

References and recommended reading

Clow, D. R., Hutchins, D. E., & Vogler, D. E. (1990). Treatment for spouse abusive males. In S. M. Stith, M. B. Williams, & K. Rosen (Eds.), *Violence hits home*. New York: Springer.

Corey, G. (1991). *Theory and practice of counseling and psychotherapy* (4th ed.). Pacific Grove, CA: Brooks/Cole. This is an excellent overview of major theories of counseling and psychotherapy, presented in a concise manner.

Ellis, A. (1982). Major systems. *Personnel and Guidance Journal*, *61*, 6–7.

Hutchins, D. E. (1979). Systematic counseling: The T–F–A model for counselor intervention. *Personnel and Guidance Journal*, *57*, 10, 529–531.

———. (1982). Ranking major counseling strategies with the TFA/Matrix system. *Personnel and Guidance Journal*, *60*, 7, 427–431.

———. (1984). Improving the counseling relationship. *Personnel and Guidance Journal*, *62*, 10, 572–575. Reprinted as one of the critical articles of the decade in Dryden, W. (Ed.). (1989). *Key issues for counseling in action* (pp. 53–62). London: Sage Publications.

Hutchins, D. E., & Mueller, R. O. (in press). *Manual for the Hutchins Behavior Inventory*. Palo Alto, CA: Consulting Psychologists Press.

Lazarus, R. S. (1984). On the primacy of cognition. *American Psychologist*, *39*, 124–129.

Mueller, R. O., & Hutchins, D. E. (in press). Behavior orientations of student residence hall counselors: An application of the TFA system. *Journal of College and University Student Housing*.

Mueller, R. O., Hutchins, D. E., & Vogler, D. E. (1990). Validity and reliability of the Hutchins Behavior Inventory: A confirmatory maximum likelihood analysis. *Journal of Measurement and Evaluation in Counseling and Development*, *22*, 203–214.

Smith, D. S. (1982). Trends in counseling and psychotherapy. *American Psychologist*, *37*, 802–809.

Tieman, A. R. (1991). *Group treatment for female incest survivors using TFA systems(™)*. Unpublished doctoral dissertation, Virginia Polytechnic Institute and State University, Blacksburg, VA.

Ward, D. E. (1983). The trend toward eclecticism and the development of comprehensive models to guide counseling and psychotherapy. *Personnel and Guidance Journal*, *67*, 154–157.

Zajonc, R. B. (1984). On the primacy of affect. *American Psychologist*, *39*, 117–123.

Giving Feedback to Clients and Receiving Feedback as a Helper

3

Key points

1. Feedback gives a client an accurate, timely observation of behavior.
2. The helper's reaction to the client's behavior (thinking, feeling, and/or acting) can be included in feedback.
3. The client can learn to give self-feedback to evaluate and monitor behavior.
4. Videotape and audiotape can be used very effectively to give feedback to a client.
5. Students learning to be helpers can benefit greatly from feedback as they use this textbook.

Feedback: What is it?

Critical to changing behavior—thoughts, feelings, or actions—is the giving and receiving of clear, accurate feedback: personal observation of and reaction to how someone else thinks, feels, or acts. Feedback entails the accurate description of behavior (sometimes with encouragement or personal reaction included with the description). In a helping relationship, feedback usually goes from helper to client. The client, however, can also learn to do accurate, timely self-feedback and to examine and evaluate aspects of personal behavior. This text includes exercises that require accurate feedback from an observer to a neophyte helper.

- HELPER (at the second session with a client): I noticed that you maintained eye contact with me today when we first met and that your handshake was firm and definite. Congratulations on good practice of two behaviors that we have already discussed! This will really help in your goal of meeting people in a positive way.

Technically, *feedback* is the squawk or squeal a listener hears when an audio system reacts to itself. It is defined by *The New American Webster Handy College Dictionary* as " . . . a return of part of the output of a system to the input." In a helping relationship, feedback occurs when the helper (or client) gives back to the client (or helper) an observation or reaction to behavior. In the words of the above definition, the helper's words of observation become the "return" of the behavior—the "output of a system"—to the "input," the client's perception of behavior.

In the helping relationship, *useful feedback* is an accurate, descriptive, nonevaluative assessment of one's behavior. It is *accurate* because it is given by an observer who attends very carefully to both what is said (verbal aspects) as well as how it is said (vocal and nonverbal aspects)—see Chapter 5, "Listening for Understanding." Useful feedback is *descriptive* because the helper, almost like a video recorder, provides a specific, detailed account of both verbal and nonverbal aspects of behavior. The client can thus get a clearer picture of how various behavior is being "received" by others. Typically, useful feedback is *nonevaluative*. Labels that carry emotionally loaded judgments ("that was stupid, idiotic, dumb, ridiculous, etc.") are avoided, which reduces the defensiveness of the client receiving feedback.

Useful feedback can, however, be coupled with a candid appraisal of the *consequences* of specified behavior for both the client and others.

- HELPER: When we first met, you avoided eye contact with me and gave a very limp, weak handshake. To many people, those two behaviors give the impression that you are neither interested nor sincere in wanting to know them.

In this feedback, the helper described what happened and then supplied information about probable consequences of the client's behavior.

Helpers must learn to give accurate feedback to their clients, to teach their clients to give self-feedback, and to request and accept feedback on their own work as helpers. Just as accurate feedback is helpful for behavior change in a client, it will also assist those learning to be helpers as they study this book. At the end of most chapters are role plays and activities. The helper is encouraged to obtain as much feedback as possible from fellow students as well as from coached clients.

Providing feedback

Feedback should cover behavior that has both positive and negative consequences. A helper who gives an insecure client—or even a psychologically secure one—information only about negative behavior adds far less to a client's learning repertoire than if the helper had also pointed out positive aspects of behavior. Building on and improving individual strengths provides a good point of growth and helps overcome negative behaviors. A mix of both positive and negative feedback is desirable. Just as too much negative can overwhelm a client, providing feedback on only positive behavior does not give an accurate picture of the client's behavior.

A model for providing feedback taught by the National Association of Secondary School Principals (NASSP) in its Leader 1 2 3 program requires that the person performing the behavior provide self-feedback first, followed by those coaching. Positive feedback is always given first, followed by suggestions for improvement. Translated for this book, the client is asked to provide self-feedback first, followed by the helper or group members. Or, the helper first indicates that feedback is desired, then provides self-feedback, and finally gets input from others.

Leader 1 2 3 (Hersey & Hakel, 1988) suggests these points for providing feedback, adapted here for a helping setting:

1. *Ascertain clearly the client's goals* before observing the behavior and providing feedback.
2. *Observe carefully*, perhaps making written notes, and *note specific behavior* that relates to the client's goals. This is especially empha-

sized in Chapters 2 and 10, which describe a systematic method of assessing behavior.

3. *Show approval* for effective use of behavior through both verbal and nonverbal methods.
4. *Demonstrate or describe more effective behavior* when poor performance occurs; continue practice of effective behaviors only. (See Chapter 12, related to stopping ineffective behavior, continuing and strengthening effective behavior, and learning new behavior.)
5. *Repeat the behavior practice* until performance is consistently acceptable.
6. *Focus on the behavior,* not on the client.
7. *Ask the client what he or she wants to improve.*
8. *Focus on one or two specific behaviors* to improve at once; give feedback on those specific behaviors.
9. Discuss and *agree on the specific actions* to be completed to improve behavior.
10. *Follow up* with the client to ensure that feedback is understood and that subsequent practice is effective, remembering that "*perfect* practice makes perfect."

Leader 1 2 3 also teaches *behavior for seeking feedback*. This behavior will be important for clients to learn to continue to improve their own behavior away from the helping setting. Seeking feedback is also important for helpers who want to continue improving their own effectiveness in working with others. Actively seeking feedback is key for persons using this text to learn to be better helpers:

1. *Decide what feedback is needed* and who can best provide it.
2. *Ask for specific feedback.* Be direct in asking about a problem or about options for behaving differently in the future.
3. *Rephrase the problem* if needed, probing to get full understanding of the feedback that has been provided.
4. *Listen openly,* not defensively, watching for nonverbal clues that may contradict open listening, even for solicited feedback. If someone gives an observation with which you do not agree, there must be some reason it was made. At least withhold an immediate judgment and *consider* the feedback. It is your decision, of course, what to do with it.
5. *Ask for ideas.*
6. *Thank* (sincerely) those who gave feedback.

Uses of feedback

Giving and receiving feedback are skills that every helper must know and teach to clients. Behavior change is the goal of helping, and accurate feedback is the means of letting the client know how he or she is changing.

1. After determining the goals of the client, the helper elicits self-feedback from the client and provides helper feedback on the current status of the client's behavior.
   - CLIENT: Whenever I begin to talk in front of a group, I stutter and have trouble breathing. I feel like I'm going to faint while I'm standing there.
2. As the client practices behavior, the helper provides feedback regarding the effectiveness of that behavior by describing the behavior and the helper's reaction to it. Members in a helping group provide important feedback to each other; all feedback need not come from the professional helper.
   - GROUP MEMBER: When you called me by name and put your hand on my arm, I felt your sensitivity to my feelings, even though you were confronting me about what I'd just said.
3. Clients can learn to critique their own behavior and provide self-feedback.
   - CLIENT: I spoke up in the group and said what I thought. Later on, though, I drifted off into my own thoughts and didn't keep myself involved in what was going on with the other group members.
4. Clients can give themselves feedback to monitor their own behavior when they are away from the helper.
   - CLIENT: I've been sitting here for thirty minutes but I've only finished three math problems and I should already be done. What have I been doing with my time—oh, I did get up to get a snack and the phone rang and . . . maybe I really haven't been sitting here all that time.

   As you can see, the client gave accurate, descriptive information on behavior that actually occurred.
5. Using videotaping or audiotaping is a good way for a client to view a behavior. The client can then give feedback on what is visible or audible, giving opportunity for reflective feedback by both the client and the helper.
   - HELPER: The video shows what you do with your glasses when you are talking in a group. Can you see how it's sort of threatening when you punch the air with your glasses and point them at Dan when he's speaking?

Giving and receiving feedback can be used in many ways throughout the helping process. It can be used to

- *Assess* a client's behavior as the helper hears the client give a self-critique and as the helper uses feedback to evaluate current status and progress of the client.

- *Establish goals* for what the client needs to do to improve behavior and become more effective.
- *Design procedures* to assist the client in making necessary changes to engage in more effective behavior.
- *Terminate* a helping relationship as the helper uses feedback to summarize progress the client has made and reminds the client of the strategy of giving self-feedback to monitor and correct future behavior outside the helping setting.

Cautions about feedback

1. Too much negative feedback can defeat the purpose of the feedback by providing an overload of negative information, defeating the client before improvement can begin.

2. Useful feedback is based on accurate observation and assessment of behavior. A client who cannot recognize effective behavior cannot be expected to give accurate feedback.

3. Feedback, especially in a helping group, can sometimes slip from the behavior to the person. The helper must be sure feedback remains focused on the behavior and not on the person. Value judgments must be avoided. Rather, give feedback on the effectiveness or appropriateness of behavior.

4. Openness in accepting feedback is a trait to be cultivated. To not be defensive about negative feedback can be difficult. An atmosphere of trust is necessary for ongoing, useful feedback to be given and received between helper and client. The helper must be skilled in deciding how much and what kind of feedback each client needs and when, by whom, and how the feedback is to be delivered.

5. Giving and receiving feedback requires practice, especially for aspiring helpers. It is a skill to be practiced before it is used with clients. In various role-playing situations in this book, there are many opportunities to practice giving and receiving *useful* feedback throughout the learning process.

6. Some clients are adept at giving themselves *negative* feedback to the extent that they elicit group sympathy instead of accurate group reaction. The client gives such strong negative self-feedback that the group or the helper jumps to the rescue and begins saying only positive things. This maneuver, done either consciously or unconsciously by the client, prevents giving and receiving accurate feedback. The helper must ensure that the client gives accurate self-feedback, both positive and negative, so that a sympathetic reaction from the helper or the group does not block delivery of accurate feedback.

7. Videotaped feedback may be very threatening for some individuals who have never viewed themselves before. The helper will want to allay anxiety and allow some time for familiarity with equipment and experiencing of videotaping before relying too heavily on this strategy for providing feedback. Typical behavior should be occurring on the videotape before

beginning to critique and offer feedback about taped behavior. To a lesser extent, some may find audiotaping inhibiting to natural behavior.

Practicing giving and receiving feedback

At the end of most chapters, suggestions for practice of the skill or strategy encourage students to invite feedback on performance. The ideas discussed in this chapter should be practiced each time you solicit feedback on performance or give feedback to a fellow student.

References and recommended reading

Cooker, P. G., & Nero, R. S. (1987). Effects of videotaped feedback on self-concept of patients in group psychotherapy. *Journal for Specialists in Group Work, 12*(3), 112–117. This article discusses use of videotaped feedback in a therapeutic group setting.

Corey, M. S., and Corey, G. (1992). *Groups: Process and practice*. Pacific Grove, CA: Brooks/Cole. The Coreys talk about the importance of accurate, timely feedback to clients.

Hersey, P. W., & Hakel, M. D. (1988). *Leader 1 2 3: A developmental program for instructional leaders*. Reston, VA: National Association of Secondary School Principals. This leadership training program for school administrators includes extensive training on giving and receiving feedback.

Ivey, A. E. (1983). *Intentional interviewing and counseling*. Pacific Grove, CA: Brooks/Cole. Chapter 2, "Attending Behavior: Basic to Communication," includes a behavior-feedback sheet.

Kruger, L. J., Cherniss, C., Maher, C. A., & Leichtman, H. M. (1988). A behavior observation system for group supervision. *Counselor Education and Supervision, 27*(4), 331–343. The authors discuss using feedback for helpers in training.

Mitchum, N. T. (1989). Increasing self-esteem in Native-American children. *Elementary School Guidance and Counseling, 23*(4), 266–271. The authors discuss using feedback in a group setting.

Morran, D. K., Robison, F. F., & Stockton, R. (1985). Feedback exchange in counseling groups: An analysis of message content and receiver acceptance as a function of leader versus member delivery, session, and valance. *Journal of Counseling Psychology, 32*, 57–67.

Building Blocks of the Effective Interview

PART TWO includes chapters that will help readers learn skills that are basic to the helping process.

TWO

Structuring the Helping Interviews

4

Key points

1. *Structure* includes the therapeutic relationship, setting, rationale, and procedures used from beginning to end in the helping process.
2. Specific elements of structure include the following:

   The physical environment
   The organizational or institutional setting
   Limits of confidentiality
   Assessment of the helper, client, and problem situation
   Expectations of the helper and client
   Roles of the helper and client
   The initial interaction between helper and client
   Goals and purposes of the helping relationship
   Procedures used within and outside the interview
   The working relationship between helper and client
   Relative emphasis on thoughts, feelings, and actions
   Emphasis on the past, present, and future
   Time dimensions
3. Although some elements of structure relate mainly to the initial interview, many elements of structure are used *throughout* the helping process.

THE ability of the helper to clearly state intentions of the helping interview is positively related to outcome assessments (Kelly, Hall & Miller, 1989). Practice responding to questions outlined in this chapter can promote confidence on the part of the helper.

Structuring the helping process

Hundreds of different approaches to the helping process are available, and they differ in terms of training of the helper, interpersonal dynamics, setting, problems, treatment goals, theory techniques, and strategies used. Because of the conceptual or theoretical rationale, these approaches also differ widely in what is emphasized—early childhood experiences; parental relationships; dreams; sexuality; family, career, pastoral, or health concerns; use of conscious and unconscious processes; emphasis on the past, present, and future; emphasis on cognition, affect, and action; and so on. A major objective of this chapter is to outline structural characteristics of helping relationships that tend to be shared across different approaches to the helping process, regardless of what they may be called in specific instances. Frank (1973, 1981) sees four common factors, or structural components, that most helping approaches share:

- A therapeutic *relationship* with clearly defined roles and expectations for helper and client

- A *setting* in which the client can receive help for emotional distress
- A *conceptual rationale* that explains the client's behavior and offers guidelines for positive change
- A set of *procedures* by which the client and helper operate

These common factors set the tone for this chapter.

The setting: Organizational, institutional, or private

Where the helper works—whether hired by a private or public organization, agency, institution, or self-employed in private practice—has implications for both helper and client. The setting has a great deal to do with what happens and how it happens in the helping process and impinges on many other aspects. As a function of the setting, the helper has different kinds of opportunities as well as limitations or restrictions under which to operate.

In schools, business, industry, agencies, and mental health or medical settings, there may be guidelines or specific policies that relate to confidentiality, the kinds of problems, behavior, or situations that must be reported to others (e.g., alcohol and chemical dependency or abuse; probation and parole violation; verbal, physical, and sexual abuse; HIV status). The helper must be familiar with guidelines of the organization as well as local, state, and national policies, regulations, and laws that relate to practice in the helping profession. It is imperative, as well, for the helper to be aware of legal and ethical guidelines that relate to the helping process and the particular setting in which one works (Corey, Corey & Callanan, 1988; Huey & Remley, 1988). Thus, there are limits to confidentiality as a function of where one works. In private practice, most behavior of the client can be regarded as confidential, subject to state and national laws as well as the professional codes under which a helper functions. However, if the helper becomes aware of violation of law or threatened personal harm to the client or others, direct action must be taken (*Tarasoff*, 1974).

The physical environment sets the tone

Depending on the work setting, there may be a wide range of conditions in which the helping relationship occurs, from a private office suite of a Licensed Professional Counselor or psychologist; a blended family or partner's apartment visited by a social worker; a major business where the client may work with an industrial psychologist, career development or employment assistance program counselor; to a comprehensive mental health center where a wide range of health professionals may work as a team. Sometimes, for example, a school counselor or helper in business or industry may have such a wide range of roles and relationships that it may be necessary—or even desirable—to operate in a relatively public setting such as a cafeteria, hallway, lounge, or other relatively open place. Different environments offer the essential confidentiality or privacy desired because of the helper's mix of roles and responsibilities.

The most frequent physical setting, however, is likely to be an office in which there is a one-to-one relationship and in which complete privacy is assured. When open, two-way communication is desired, a more relaxed, informal arrangement is structured by where the helper and client sit.

Often, in a private office, some variation of the setting described below is used, with both the helper and client sitting in a comfortable arrangement that does not suggest the helper is more powerful than the client. If a desk is involved, the client often sits at the side of the desk rather than directly in front of the desk—a situation that implies differential power and authority. A variation on power also exists when working with young children, individually and in groups. A helper who works with children often arranges the physical environment so that the children are about on the same level, sitting in lower chairs or on a rug or pillows on the floor (Cole & Earthman, 1990).

The setting suggests *accommodating the client's concern for privacy* in order to conduct the business of the relationship. Some problems need to be addressed in a confidential manner in a place where others will not see that the interview is taking place or overhear highly confidential personal information.

- EXAMPLE: In a business setting dealing with financial counseling, the professional staff had moved into new facilities with five-foot high partitions around small areas reserved for conferences. The director and staff noted an immediate reduction of talk about personal financial concerns. Also, whenever people came into this new office arrangement, they would be able to see who went where . . . and surmise why. Because many clients as well as staff expressed concern about the lack of privacy, this trendy new office facility was physically changed to allow clients the comfort they needed to discuss their personal concerns with the helper.[1]

Courtesy to the client is something that should be given primary attention. Personal privacy and confidentiality should be regarded as a high priority. This is important whether working with couples, families, or children in subgroups or with industrial and health groups, among others. Minimizing interruptions, ringing phones or talking on phones, and other preventable interruptions helps the client to know that the helper's full attention is focused on the individual. Sometimes it is essential to plan sessions early (or help the client plan) for inclusion of other family members, coordinating the work schedule and transportation, and so on in structuring a helping relationship.

Structuring the relationship

There are several important aspects to consider in the *initial phase* of a helping relationship. Depending on the setting, some questions from the

[1] Thanks to Ed Spencer.

client can be anticipated, printed, and made easily available for the client to read. General questions about the helping process, what goes on in the relationship, procedures, and expectations can be effectively handled through printed information. This is true in a variety of settings ranging from schools through health, business, and industry to private practice work.

One of the leading writers in the area of committing the structure of the helping process to print is Weinrach (1989). He advocates basing written guidelines on issues most frequently raised by clients as well as including areas where potential conflicts may exist (Weinrach, 1989, p. 299):

- How often can I expect to have an appointment?
- How might I reach you if I feel I need to?
- What happens if I forget an appointment?
- How confidential are therapy sessions?
- What do I do in the event of an emergency?
- When is it time to end treatment?
- What are my financial responsibilitics?
- How often do I obtain reimbursement from my insurance company?

Guidelines should be viewed as flexible, as a discussion device before problems occur, and as a vehicle that encourages both client and helper to deal openly with important issues such as progress in the helping process (Weinrach, 1989). Whether written or discussed, such concerns may become issues in various helping relationships, and the helper is well advised to anticipate and attend to these. Structuring such concerns in writing makes for effective and efficient use of time besides stimulating open discussion about a variety of concerns before they become problems.

Ivey and Matthews (1984) propose a five-stage meta-model, which is outlined below, for structuring clinical interviews. Chapters in this book that address these concerns are indicated in parentheses.

1. *Rapport and structure:* meeting the client at his or her own model of the world (Chapters 2 and 3)
2. *Gathering information:* defining the problem, identifying assets, and carefully attending to client behavior (Chapters 2–10)
3. *Determining outcomes:* where does the client want to go, and what do you want to have happen? (Chapter 11)
4. *Exploring alternatives* and confronting client incongruities: what are we going to do about it? (Chapters 11 and 12)
5. Generalization and transfer of learning: will you do it? (Chapters 12 and 13)

Assessing the helper, client, and problem

An evaluative screening may start even before the helper sees the client—as in the case of a telephone referral or a review of information on paper. Typically, the helper makes an initial assessment during the first meeting with the client.

From the initial greeting and throughout the interview, the helper assesses the appropriateness of working with each client.

As a helper, it is important that you think through the kinds of things you expect to happen in the helping process. Advance thinking about your approach lets you respond to clients' questions with clarity and confidence. The following questions offer a starting point:

1. Am I *personally* and *professionally qualified* to work with this client who has this particular concern or problem in this specific situation?
2. Do I understand the unique personal, educational, social, and cultural aspects of this client enough to be able to assist in this situation?
3. Should the client be *referred* to a helping professional who has more or different specialized training and skills—such as a licensed counselor, psychologist, social worker, psychiatrist, marriage and family therapist, substance-abuse specialist, career or pastoral counselor, health and medical personnel, or others?
4. What is my role as a helper in the relationship?
5. What kinds of things do I see as important variables in the helping process?
6. What kind of behavior (thoughts, feelings, and actions) do I expect of the client both in and outside the interview setting?
7. What kind of commitment do I expect of the client in terms of time, work, and responsibility?
8. What about confidentiality in the setting in which I work?
9. Am I prepared to answer questions about the helping process, such as those outlined earlier by Weinrach?
10. What legal, ethical, and moral considerations must be considered before working with this client? Am I knowledgeable and comfortable working under these conditions?
11. Would I be entering into a dual relationship that might compromise the client or myself because of other interactions with the client outside the helping relationship?

If the helper is working under the supervision of another professional, questions such as these might best be handled cooperatively, especially if there are concerns about the training and ability of the helper to assist the client. If such decisions are not made prior to the first interview, they must be made at an appropriate time, recognizing that some potential problems may not emerge well into a helping relationship.

Client expectations

What does the client expect to happen during this process? Books, magazines, radio, television, and video programs have all illustrated aspects of the helping process in different ways. Some portrayals of caseworkers, counseling, therapy, and psychological and psychiatric interventions have shown the helping relationship in ways that may have little to do with how you work. Does the client expect to lie down on a couch and free-associate or talk about anything that comes to mind? Talk mainly about feelings and emotions, dreams, cognitively oriented problem solving, or what?

The client may have had a friend, family member, or acquaintance who has seen a professional helper who has worked differently from you. In any case, it is essential to let the client know your expectations at appropriate times in the interview as well as to solicit the client's expectations about what will happen in the helping process.

Is the client prepared to accept responsibility for personal behavior in the helping process? No change can be made without a realistic, sincere commitment from the client. Changing personal behavior means doing whatever it takes to think, feel, and act differently. This cannot be done unless the client assumes major responsibility for changes during and outside the helping interview and makes a commitment to do what is necessary (commitment is discussed more fully in Chapter 11).

- HELPER: To do an effective job of exploring your career options, you will be doing several things in the helping process. These include talking to other people about their work, reading from sources in the career library, taking tests related to work and leisure interests and aptitudes, as well as thinking about and completing a structured workbook about yourself. This process will take about four to six weeks. Are you willing to commit the time and energy to do these things?

Structuring by the helper, such as illustrated above, might lay out a process that is more involved and time-consuming than the client had thought when visualizing a quick fifty-minute fix that would lead immediately to the next job. Thus, before much time has elapsed, the helper is wise to assess the client's commitment to do what is necessary both in and outside the helping setting.

What are the client's expectations in terms of time (for each interview), length of the helping process (beginning to termination), frequency of sessions, payment, and other aspects? Also, what about the unique personal, educational, social, and cultural aspects of the client? Clients from other cultures may have radically different expectations about what is involved in the helping process. The helper must be aware of client expectations if the helping process is to succeed. The greater the difference between expectations and reality, the more likelihood of failure. Stated in a positive way, if the client's expectations of what will happen are accurate and realistic, there is a greater probability of success (Hoppock, 1976).

Many helpers work in settings such as social service agencies, chemical-dependency treatment centers, churches, hospitals, or rehabilitation centers in which they are exempt from licensure or training other than what the institution provides. What the helper does in such settings is regulated by the organization. In each case, the helper must be aware of specific local, state, and regional requirements that relate to the helping process. In addition, professional standards of training, care, and conduct need to be considered in making judgments that are in the best interests of the client. (See "References and recommended reading" related to legal and ethical issues for further information on this topic.)

The initial structure provided by the helper should not be an encyclopedia of information or questions rattled off endlessly. Rather, the helper should respond directly and matter of factly to questions raised by the client. Practice responding to questions such as those in this chapter will help you respond appropriately when they arise during the helping process.

The first few minutes are important.

How the counselor greets the client is the client's first clue to whether the helping process will be successful. The greeting should be personable yet businesslike, and the client should be made to feel welcome and relatively at ease.

- HELPER: Hello. I'm [give your name and shake hands, if appropriate]. Please have a chair [perhaps signaling where the client should sit]. Tell me what brings you here.

What about the tone of the interview—is it businesslike, too relaxed, or overly serious? Both parties are here for important business that must be accomplished in a limited time period. In general, the sooner the helper and client get down to the business of helping, the more helping business will get done. There is a place for some exchange of information and for *appropriate, genuine, mutually shared situational humor* at various points in the interview. However, as a general rule, talk unrelated to the process of helping the client is inappropriate.

What is the client's mood (e.g., relaxed, anxious, depressed, hostile, skeptical)? How can the counselor respond appropriately?

- HELPER: You seem to be [short of breath/upset/nervous/or otherwise acknowledging the client's nonverbal behavior]. How can I help you?

Sometimes cues such as those above are very strongly related to the client's concern and are best addressed immediately. When the helper is alert to both verbal and nonverbal cues and responds appropriately, the client is most likely to sense that the helper can be of assistance.

One of the first things the helper does in the initial interaction is to determine reasons the client is seeking assistance. This may vary, however, with the particular setting in which one is involved—for example, home visitations, required observations by an agency representative, and the like. In any case, the helper who anticipates and responds to concerns directly and without hesitation is likely to be perceived as one the client can trust.

- HELPER: We have about twenty minutes today before I have to catch a plane. It would help me if you could tell me why you are seeking assistance and what your needs are.
- HELPER: I see the helping process as a time when you are actively involved in working to resolve personal concerns that are troubling you. I expect that you will be able to talk openly about your thoughts and feelings, as well as what you do. Together, we will work to help you take action to resolve concerns you have.
- HELPER: Let's start with your story [reasons you are seeking help/why you came here/what your situation is].

Such statements give the client information about what the helper expects, clearly and unambiguously. There is a strong positive relationship between helpers who are able to clearly state their intentions and ratings of competence by clients (Kelly, Hall & Miller, 1989).

Procedures used in and outside the helping interview

Throughout the helping process, structuring also involves letting the client know what is likely to occur both in and outside the helping interview. Naturally, these activities are a function of the setting in which the process occurs. The helper discloses information about procedures to the client as appropriate or when asked.

- HELPER: You asked what would be involved in order to work on relating more effectively with others. Basically, what we'll do here is to take a close look at what you currently do, probably through role playing. Then we'll assess strengths and limitations of this and make a plan for how you can improve. You'll likely be practicing some new ways of interacting with me right here. Then we'll work out some further practice you can do on your own. If you really work at these things, you should begin to see progress in the next two to three weeks.

Some aspects of structure are implicit—a function of what the helper does (such as what the helper focuses on in the interview—client thoughts, feelings, and actions; dreams; interpersonal concerns, etc.). Other aspects may be more explicit, such as verbally outlining specific kinds of actions that may occur both in and outside the interview setting.

In the actual interview, besides the relationship skills presented in following chapters, the helper uses a host of procedures designed to help the client achieve goals (Chapters 11 and 12). These may include role playing, behavior rehearsal, imitation, imaging or fantasizing, relaxation training, gestalt techniques, sculpting, testing and written activities, and so on. Many procedures and strategies might be used in the process of helping the client (see Part Four). Such procedures are often associated with the helper's theoretical preferences obtained by training and experience.

Outside the helping interview, there are thousands of behaviors the client might be asked to engage in to assist in learning skills to achieve larger goals. The client may be asked to do things at home, at work, with friends, in private or in public settings. A brief sample of the possibilities includes reading, interviewing, talking to others, learning new skills and activities, taking tests, keeping records, starting a diary or journal, charting and recording progress on specified tasks, listening to audiotapes and videotapes, and practicing a variety of behavior (see Chapter 11 and Part Four).

Elements of structure include explaining what kinds of events can be expected to occur during the process of helping, from the initial interview through termination and follow-up. Some aspects of structure will occur mainly in the initial phases of the helping process (the initial greeting, discussion of time constraints, roles, confidentiality, potential action, etc.). Other aspects of structure may take place throughout the helping process (such as clarification of expectations and actions both in and outside the interview setting). The helping process works best when the helper is explicit and straightforward with the client, describing what will be happening and why. In a process where success depends so much on active client participation, success is enhanced by accurate client knowledge.

Summary

Structure relates to:

1. The therapeutic relationship, setting, rationale, and procedures used in the helping process.
2. The roles, goals, actions, procedures, and time dimensions used *throughout* the helping process from start to finish.
3. Confidentiality between the helper and client.
4. Unique aspects of each client's behavior, concerns or problems, specific situation, as well as social and cultural background.
5. The wide variations of settings in which the helper works.
6. Expectations of both client and helper.
7. The emphasis on thoughts, feelings, and actions.
8. Actions both in and outside the interview setting.

Practice

1. Find a person who agrees to be your coached client (to role-play a client). Assume that you and your coached client have just met for the first time.
2. Choose one situation from the list of role plays that follows these instructions. Before starting to role-play, clarify specifics of the situa-

tion: the client's age, gender, the setting of the interview, and other details necessary for the interview.

3. Read through the "Focus Checklist," which follows the role-playing situations. Items on the checklist provide some guidelines for structuring your mini-interview.
4. Conduct a three- to five-minute mini-interview for *each* role-playing situation. For these first exercises, assume that you are meeting the client for the first time so you can practice greeting the client, exchanging names, making the client comfortable, and so on. Practice as many different structuring techniques as you can. Remind your coached client to keep responses fairly short because *you* are the one who is practicing your skills. *Note*: Even though some aspects of structuring will seem rushed or crammed when you do them for this exercise, practicing saying different things related to structure will help you later. If you say the words *now,* you will be comfortable using them when it is more appropriate and as you progress through other helping skills. Also, because you will use only a few aspects of structuring in any single mini-interview, practice a number of different role-playing situations, saying different things until you become comfortable with many different aspects.
5. Record your interaction (or have an observer present or videotape the interaction).
6. Following the mini-interview, give yourself feedback, ask your coached client for feedback about the interaction, and, especially, get information related to your structuring.
7. Listen to the recorded interview and answer questions in the "Focus Checklist," noting where you need more practice.

Role play

Remember to specify the situation so that you and your coached client know enough to get started (setting, names, age, gender, problem situation, etc.). Encourage your coached client to overdo relevant feelings and emotions so that *you* get practice responding to this dimension of behavior.

If you find you need additional role-playing situations or different ones that relate more to your particular setting, create your own. You'll get additional experience if you vary the client's age, gender, culture, concern, and setting.

1. Your client is having trouble with a child's behavior around the home and in school. This parent is at a loss for what to do to change the child's behavior and feels the situation is growing increasingly out of control.
2. Your client is a teenager. You choose the culture and socioeconomic background. The client has a *B* average as a high school senior, enjoys doing many different things, but has no specific goals. The client

feels a lot of pressure from a parent about getting into college but is confused and undecided about what to do.

3. Your client is between thirty-five and fifty-five years of age and expresses the following concern: "We're in such deep financial trouble that I don't know how we're going to get out of this one. I have to keep two jobs just to get by. On top of that, I hate my daytime job and have been thinking about quitting."
4. Your client comes in to see you about making a job change. You notice the client greeted you in a very awkward manner and seems to be very nervous and uncomfortable talking with you in the interview.

Focus checklist

Use the following guidelines to assess how you used structure in the mini-interviews.

1. Was the physical setting appropriate for the situation?
2. Did you exchange names?
3. Was there a handshake or friendly greeting?
4. Did you make the client feel comfortable?
5. Was there a minimum of social chit-chat?
6. Did you discuss confidentiality?
7. Did you discuss the helping process as it relates to the client's concern?
8. Did you discuss your expectations?
9. Did you discuss the client's expectations?
10. Did you discuss the time dimension?
11. Did you discuss potential procedures that relate to the kinds of actions that might be done in the interview setting? Outside the interview setting?
12. Is the client's concern or problem within your scope of competencies and the setting?
13. Did you discuss costs of the helping relationship?
14. What other aspects of structure could have been discussed? (The answer to this should suggest tasks for you to practice in some of your next mini-interviews.)

References and recommended reading

Cole, C. G., & Earthman, G. (1990). Physical environment and the school counselor. *Educational Facility Planner*, *4*, 20–22.

Corey, G., Corey, M. S., & Callanan, P. (1988). *Issues and ethics in the helping professions* (3rd ed.). Pacific Grove, CA: Brooks/Cole.

Frank, J. D. (1973). *Persuasion and healing* (2nd ed.). Baltimore: Johns Hopkins Press.

———. (1981). Therapeutic components shared by all psychotherapies. In J. H. Harvey & M. M. Parks (Eds.), *Psychotherapy research and behavior change*. Washington, DC: American Psychological Association.

Hoppock, R. (1976). *Occupational information* (4th ed.). New York: McGraw-Hill.

Huey, W. C., & Remley, T. P., Jr. (1988). *Ethical and legal issues in school counseling*. Washington, DC: American School Counselor Association.

Hummel, D. L., Talbutt, L. C., & Alexander, M. D. (1985). *Law and ethics in counseling*. New York: Van Nostrand Reinhold. A comprehensive book for both practitioners and students. Laws and court cases are discussed along with practical application.

Ivey, A. E., & Matthews, W. J. (1984). A meta-model for structuring the clinical interview. *Journal of Counseling and Development*, *63*, 237–243.

Kelly, K. R., Hall, A. S., & Miller, K. L. (1989). Relation of counselor intention and anxiety to brief counseling outcome. *Journal of Counseling Psychology*, *36*, 158–162.

Tarasoff v. *Regents of the University of California*, 529 P 2d 553 (Calif. 1974).

Weinrach, S. G. (1987). The preparation and use of guidelines for the education of clients about the therapeutic process. *Psychotherapy in Private Practice*, *5*(4), 71–83.

———. (1989). Guidelines for clients of private practitioners: Committing the structure to print. *Journal of Counseling and Development*, *67*, 299–300.

Listening for Understanding

5

Key points

1. Active listening and attending to nonverbal behavior enables the helper to understand the client's meaning.
2. The helper attends to what and how the client says things, both verbally and nonverbally.
3. Body language (e.g., eye contact, facial expressions, gestures, and physical aspects of behavior) assists in interpreting the client's behavior.
4. The helper must be aware of the client's background and culture to interpret behavior accurately.
5. The helper can better understand the client by combining the client's verbal and nonverbal behavior with an accurate awareness of the client's personal, social, educational, spiritual, and cultural background.
6. Similarities and discrepancies among the client's thoughts, feelings, and actions provide clues to understanding the client.
7. Nonverbal behavior is a two-way street because the behaviors of the client and helper may communicate different messages to each other.

IF you observe people talking in small groups at informal social events you may see several of the following:

- People talk in rapid succession, one person quickly following another.
- Each person contributes his or her own two-cents worth of conversation.
- There are practically no periods of silence. If too much silence occurs, individuals move to other groups in search of more active or more interesting conversation.
- Comments and questions frequently do not relate to the subject being discussed.
- People interrupt others who are speaking.
- Frequent shifts in content occur, depending on who is talking.
- The continuity of the topic falters from one person to another.
- More than one person speaks at a time; perhaps several speak at the same time.
- One person may dominate conversation, giving others little opportunity for input.

Each of these actions indicates poor listening and attending to what is being said and how it is said. In many interactions between people in social settings, informal encounters, and work settings, people do a poor job of listening. One authority estimates that in business and industry billions of dollars are lost each year because of the "numerous listening mistakes" people make (Steil, 1980, p. 65). There is no room for listening mistakes on the part of the helper in the helping relationship. The good helper listens carefully to what the client says and how the client says it—both verbally and nonverbally.

Active listening in the helping relationship

In the helping relationship, the helper promotes good interaction by focusing attention on and actively listening to the client. Steil suggests that good (active) listening is more than hearing what is said: "There are three other important components: interpretation, which leads to understanding or misunderstanding; evaluation, which involves weighing the information and deciding how to use it; and finally, responding, based on what was heard, understood, and evaluated" (1980, p. 65). In attending to the client, the effective helper usually

- Listens carefully to what the client says and how it is said.
- Avoids interrupting and allows the client to complete sentences and ideas.
- Uses silence to encourage the client to continue talking, thus giving the client time (and space) to verbalize thoughts and feelings that may be difficult to talk about (especially if the client has never before expressed them to another person); see Chapter 8.
- Uses reflection to clarify the client's meaning. Reflection is mirroring the essence of the client's communication in concise terms (see Chapter 7).
- Asks questions to ascertain important details, including answers to the questions of what, where, when, how, under what circumstances, how often, and the like; see Chapter 6.
- Helps the client systematically explore relevant content.
- Reflects similarities and discrepancies among what the client says (thoughts), how the client says it (feelings), and what the client does (actions).
- Elicits feedback from the client to determine and ensure accuracy of the helper's perceptions of the client; see Chapter 2.

Active listening is the basis on which every other aspect of the helping relationship is built. Nothing else can proceed effectively if there has not been active listening, during which the helper has understood the client.

Nonverbal aspects of behavior

A critical link exists between what is said (or not said) and how it is said (or not said). *Nonverbal behavior* primarily refers to behavior other than the specific words the person uses. Words relate to *what* is said. Nearly everything else relates to *how* it is said. Mehrabian (1971) says that most people pick up only about 7 percent of a message from *what* people say, 38 percent from vocal aspects of their communication (tone of voice, volume, etc., but not the actual words used), and 55 percent from other nonverbal aspects of their communication (e.g., body language).

Specific aspects of nonverbal behavior include

- Vocal or subvocal sounds.
- Vocal tones, or pitch.
- Speed of talk.
- Body language, including eye contact, facial expressions, gestures, posture, and overall body movement.

One's body language and nonverbal behavior give especially useful cues to that person's feelings—including moods, mental states, and physical comfort or discomfort—that provide information about the client's behavior that can be investigated more specifically through talk. As examples, consider the following:

- A strained tone and guttural sounds could indicate either a certain reluctance to express oneself or a sore throat.
- Very slow, deliberate, sometimes stammering speech may indicate depression, anxiety, nervousness, fear, uncertainty, or disability. In contrast, more rapid, higher-pitched, spontaneous talk may convey the client's nervousness or enthusiasm for a subject.
- Folding of one's arms might indicate a person who does not wish to interact with others, who is physically cold, or who has not learned to express thoughts verbally (Zimbardo, 1977).
- Evading eye contact may indicate lack of confidence or avoiding the topic or the helper. It could also mean that eye contact is not usually given in that person's culture or that the client is merely being respectful of an authority figure or an older person.

The effective helper notes similarities and incongruities between verbal and nonverbal content—eye contact, facial expressions, gestures, overall body language, energy level, and the like. Discrepancies in perceptions may indicate differences in background between client and helper.

Influence of cultural background

The helper must carefully consider factors that may affect both the verbal and nonverbal behavior of any person. Because personal, familial, cultural, and social background can have a major effect on how one behaves, such factors have to be considered when making any judgments about the meaning of nonverbal behavior. In one person, lack of eye contact may indicate avoidance behavior. In another, the same behavior may indicate that the person is listening to what is being said but comes from a culture where eye contact is not part of the group behavior (e.g., in certain Indian or Appalachian cultures). Have you ever tried to "read" a person's intentions when the individual is wearing a pair of very dark glasses so that you cannot see the eyes? If you put a great deal of emphasis on eye contact in communicating with others, this can be an unnerving experience. Talking face-to-face may be a mark of interest and concern in some cultures or a sign of disrespect in others.

Many of us are familiar with different ways individuals greet each other. In some cultures, hugs and kisses are exchanged as a greeting. In others, this same behavior may make people feel extremely uncomfortable. Initial impressions of a client's nonverbal behavior should be considered tentative hypotheses to be tested, in light of the client's personal, social, and cultural background.

How the client expresses feelings

Cultures and subcultures vary considerably in how individuals express feelings in many social settings. Many individuals do not distinguish between thoughts and feelings. They use the word *feeling,* but what they mean relates to thoughts or ideas.

- CLIENT: I *feel* I should choose this one. [Meaning: "I think (or believe) I should choose this one."]
- CLIENT: How do you *feel* about the new policy? [Meaning: "What do you think about the new policy?"]

Feelings relate to the emotional, affective, highly personal meaning we associate with different things. As Gaylin states, "Feelings are mushy, difficult, non-palpable, slippery things even by definition. In that sense they are immune to the kind of analysis to which most behavior is typically exposed. They are difficult to quantify, difficult to communicate, difficult even to distinguish within ourselves" (1979, p. 10).

It is a challenge for individuals to understand their own feelings, let alone to describe them to others. The same event can bring with it radically different kinds of emotional responses. The stirring music of the wedding march at a marriage ceremony may bring feelings of joy, relief, sorrow, happiness, or release for different persons attending the ceremony. Winning major contests (e.g., at the Olympics) or succeeding at really important tasks may bring feelings of exhilaration, release, relief, defiance, joy, or pride. Such feelings may be accompanied by a variety of actions including jumping for joy, near physical collapse, overt signs of assertiveness, hugs, tears, and other overt expressions of inner emotions.

Feelings are often difficult to communicate to others. Table 5.1 shows a few words that express anger, conflict, fear, happiness, and sadness. These words are arranged in a continuum from low to high intensity. The word sequence in the table should be regarded as only a rough guide, however, because the meaning of any word can vary greatly in a specific situation.

Perceiving relationships among thoughts, feelings, and actions

The helper must be aware of the congruence or discrepancy between what is said and how it is said. Congruence lends support for the observation that the

TABLE 5.1 Words That Express Feelings

Relative Intensity of Words	Feeling Category				
	Anger	**Conflict**	**Fear**	**Happiness**	**Sadness**
Mild Feeling	Annoyed Bothered Bugged Irked Irritated Peeved Ticked	Blocked Bound Caught Caught in a bind Pulled	Apprehensive Concerned Tense Tight Uneasy	Amused Anticipating Comfortable Confident Contented Glad Pleased Relieved	Apathetic Bored Confused Disappointed Discontented Mixed up Resigned Unsure
Moderate Feeling	Disgusted Hacked Harassed Mad Provoked Put upon Resentful Set up Spiteful Used	Locked Pressured Torn	Afraid Alarmed Anxious Fearful Frightened Shook Threatened Worried	Delighted Eager Happy Hopeful Joyful Surprised Up	Abandoned Burdened Discouraged Distressed Down Drained Empty Hurt Lonely Lost Sad Unhappy Weighted
Intense Feeling	Angry Boiled Burned Contemptful Enraged Fuming Furious Hateful Hot Infuriated Pissed Smoldering Steamed	Ripped Wrenched	Desperate Overwhelmed Panicky Petrified Scared Terrified Terror-stricken Tortured	Bursting Ecstatic Elated Enthusiastic Enthralled Excited Free Fulfilled Moved Proud Terrific Thrilled Turned on	Anguished Crushed Deadened Depressed Despairing Helpless Hopeless Humiliated Miserable Overwhelmed Smothered Tortured

NOTE: The context in which words such as these are used may result in shifting their intensity as well as changing the category in which they are used. Words are listed here only to suggest the range of options available to the helper seeking to identify feelings of the client. Thanks to Nanette Rollins for the format for this table.

client *means* the same thing as what was said. Discrepancies sometimes indicate differences among what the client says (thoughts), how the client expresses emotions (feelings), and what the client does (actions). The following is a brief discussion of these differences:

1. *Discrepancy between thoughts and feelings*. The client thinks one thing but feels emotionally uneasy.

- EXAMPLE: This client, who thinks that being passed over for a promotion was unfair treatment, boasts about plans to "tell off" the supervisor but feels extremely uneasy about the prospect of talking about the situation with the supervisor.

2. *Discrepancy between thoughts and actions.* The client thinks one thing but does another.
   - EXAMPLE: A child believes that people should be treated fairly, yet when involved with others in a select peer group, the client verbally insults a child of different gender.

     A person expresses the idea of sharing equally, but refuses to share (tools, toys, games, or resources) with others.
3. *Discrepancy between feelings and actions.*
   - EXAMPLE: A person is very angry at a close friend, but when the friend stops by unexpectedly to visit the person appears to be friendly and acts as if nothing is wrong.
4. *Discrepancy between feelings and actions and thoughts.* The client may have acted on impulse (feelings), but after thinking about it, wishes the action undone.
   - EXAMPLE: The client and some friends decided to get even with their high school principal, who had suspended them. The students got some cans of yellow spray paint and painted the principal's car tires. Shortly thereafter, the client began to have second thoughts and felt guilty about what the group had done.

Discrepancies or incongruencies among thoughts, feelings, and actions often indicate areas that need to be explored in more detail. The helper's awareness of these relationships—or lack of them—often provides the basis for systematic questioning that can lead to the client gaining insight.

Examples of listening to the client's thoughts and feelings

The following exercises will help you learn to be more sensitive to the client's thoughts and feelings:

* • CLIENT: Oh, I don't know . . . it really doesn't make a big difference to me what happens.

- HELPER: Somehow the expression on your face tells me that you really do care a lot more than you say about what happens.

Comment: Here, the helper perceives what the client has been talking about as well as the client's nonverbal facial expressions, which seem to be discrepant.[1]

[1] The exercise marked with an asterisk is adapted from *Essential Interviewing,* by D. R. Evans et al., and is used by permission. We strongly recommend their programmed approach for those who wish to see more examples and engage in additional practice of helping skills.

- CLIENT: I fainted in the bank about four years ago, and when I came to, a lot of people were standing over me.
- HELPER: You must have felt embarrassed when that happened.

Comment: The helper ascertains the client's feelings about the incident. The helper chose not to question the client about how or when the incident occurred.

- CLIENT (looking down at the floor): Well, I . . . started having . . . just a few drinks.
- HELPER: You find it difficult to tell me about it.

Comment: The helper attends to the client's nonverbal behavior (looking down, pausing, and talking slowly), which suggests hesitance to talk about drinking. The helper chose not to shy away from a difficult issue for the client.

The helper, too, must be aware of nonverbal cues *sent* to the client. For example, like the client who delivers contradictory messages, the helper who verbally expresses a need for openness in the helping relationship may relay a different meaning by sitting in a position turned away from the client or by avoiding eye contact.

Good communication is the foundation on which the helping process is built. The essential skills that will facilitate behavior change are active listening; attending to nonverbal and verbal behavior; noting discrepancies among thoughts, feelings, and actions; and being aware of your own communication patterns.

Summary

1. Active listening is attending to *what* the client says and *how* the client says it.
2. Attending to the client's nonverbal cues such as eye contact, facial expressions, gestures, and physical aspects of behavior can bring about understanding of the client's meaning.
3. The client's verbal and nonverbal behavior, along with sociocultural awareness, assists the helper in making tentative hypotheses about the individual.
4. Comparing verbal and nonverbal behavior helps in understanding similarities and discrepancies among the client's thoughts, feelings, and actions.
5. Some understanding of the client's world and ways of self-expression is a prerequisite to an effective helping relationship.
6. The *helper* must be aware of what is being communicated to the client by both verbal and nonverbal messages.

Practice

1. Find a person who agrees to help you practice active listening and attending to nonverbal behavior.
2. Select one situation from the list of role plays, which follows these instructions. Before starting your mini-interviews, clarify specifics of the situation: the client's age, gender, and social and cultural background; setting of the interview; and any other details essential to getting started.
3. Conduct a five-minute mini-interview for each role-playing situation. Record your interactions or have another person observe you, especially noting aspects related to your active listening, attending, and verbal and nonverbal interaction with the client. At this stage, try to gather as much information as possible about the client. Do *not* try to "solve" the client's problem.
4. Following each mini-interview, ask your coached client for feedback about the interaction. Then listen to the tape or talk to your observer and answer the questions in the "Focus Checklist," which follows the role-playing situations.
5. When you have finished your mini-interviews, try making observations concerning interactions of people around you. Instructions for the observation exercises follow the "Focus Checklist."

Role play

Suggestion: Adapt these or other situations to include coached clients who may come from different social and cultural areas than you.

1. The client appears to be wanting to say something to you but hesitates, stammers, and has a great deal of trouble expressing concerns. Among other matters, the client is concerned about potentially embarrassing personal disclosures and whether it is safe to talk about these things with you. (*Note*: This situation also provides a good opportunity for you to review some aspects of confidentiality.)
2. Your client is obviously upset with the way things have been going on the job. Very little seems to be right—the boss seems always critical, and other employees seem unfriendly and unhelpful. Others don't seem to understand what it is like to come from a culture so different from this one. Your client dreads going to work in the morning.
3. Your client constantly talks about how wonderfully everything is going, yet you know that this client is having serious difficulties in relationships with others (spouse, employer, parents, children).
4. This client is really feeling low because of grades received during the first academic term. Because of these low reports, your client, who had future events fairly well planned, has become uncertain about whether to continue in school.

5. This client's ego knows no bounds. Despite others' comments to the contrary, your client cannot accept the possibility of having a personality problem. However, this person, who has many acquaintances and is involved in diverse activities, doesn't seem to have any really good, meaningful, or lasting relationships with others.

Focus checklist

1. What nonverbal behavior gave you clues about the client's thoughts and feelings?
2. What did the client's speech patterns (hesitations, repetitions, stammering, and guttural sounds) convey to you?
3. Is there a close relationship between what was said and how it was said? If there are differences between what the client said and how it was said, what specific aspects of the client's behavior accounted for them?
4. What are similarities and discrepancies among the client's thoughts, feelings, and actions?
5. Are there social or cultural differences that may have accounted for the client's behavior? If so, what are they? Did you ask about aspects of client behavior that were unfamiliar to you?
6. Ask your coached client about your behavior. What aspects (especially nonverbal aspects) of your behavior did the client notice? Were some of the things you did distracting to the client? If so, which ones?
7. What did you do to better understand the *meaning* of the situation for the client?

Observation exercises

The purpose of this exercise is to sharpen your awareness of specific behavior that leads to your hypothesis of how the client feels. You can make observations concerning people's interactions at home, with friends, in business and professional offices, in restaurants, or on the bus, train, or plane. A variation is to watch television without turning on the sound. Then try to guess what was said and how it was said simply by observing the *nonverbal* behavior of the actors (soap operas are especially good for this).

Record your observations on a chart with four headings that looks like the top of Table 5.2. Enter information into your chart in the following manner:

Column 1, "Label": Record the emotional tone of the person you have observed. Use a word or brief phrase that describes the person's behavior. Terms you might use include *excited*, *nervous*, *sad*, *happy*, or *apprehensive*.

Column 2, "What Was Said": Write the specific words the person used.

Column 3, "How It Was Said": This indicates nonverbal and vocal

TABLE 5.2 Behavior Observation Chart

(1) **Label**	(2) **What Was Said**	(3) **How It Was Said**	(4) **Nonverbal Aspects of Behavior**
"Enthusiastic"	I *really like* teaching	With emphasis on "really liking"	Sincere tone of voice and direct eye contact with people

aspects of the client's communication. Write the *how* of what is said, noting speed, tone, pitch, and mood of the client as reflected in how the person talks.

Column 4, "Nonverbal Aspects of Behavior": Write descriptions of nonverbal behavior such as type of eye contact, facial expression, gestures, leg and foot movements, and posture. You will not always write something in every column. Possible combinations include the following:

1. Use columns 1, 2, 3, and 4 if you can clearly hear the client talking.
2. Use columns 1, 3, and 4 if you cannot hear the words clearly. In this case, pay special attention to how the client acts: the rapidity and tone of speech, posture, gestures, body position, eye contact, and facial expression. Or, use this while watching brief television episodes without the sound.
3. Use columns 3 and 4 to specifically try and guess the kind of interaction that is taking place between the clients you are observing in a group. Try to assess the emotions and feeling between individuals by observing eye contact, gestures, body posture, and movements. Be specific in describing what they do that causes you to interpret their interaction the way you do.

References and recommended reading

Egan, G. (1990). *The skilled helper* (4th ed.). Pacific Grove, CA: Brooks/Cole.

Evans, D. R., Hearn, M. T., Uhleman, M. R., & Ivey, A. E. (1989). *Essential interviewing.* (3rd ed.). Pacific Grove, CA: Brooks/Cole.

Gaylin, W. (1979). *Feelings: Our vital signs.* New York: Ballantine Books.

Gendlin, E. G. (1981). *Focusing.* New York: Bantam Books.

Hill, C. E., & Stephany, A. (1990). Relation of nonverbal behavior to client reaction. *Journal of Counseling Psychology, 37,* 22–26.

Ivey, A., & Authier, J. (1978). *Microcounseling: Innovations in interviewing, counseling, psychotherapy, and psychoeducation* (2nd ed.). Springfield, IL: Thomas.

Marsella, T., & Pederson, P. (Eds.). (1981). *Cross-cultural counseling and psychotherapy.* Elmsford, NY: Pergamon Press.

Mehrabian, A. (1971). *Silent messages.* Belmont, CA: Wadsworth.

Steil, L. K. (1980, May 26). Secrets of being a better listener. *U.S. News and World Report,* 65–66.

Sue, D. (1981). *Counseling the culturally different.* New York: Wiley.

Zimbardo, P. G. (1977). *Shyness.* Reading, MA: Addison-Wesley.

6 Asking Appropriate Questions and Systematic Inquiry

Key points

1. Effective questioning is one of the most useful ways of understanding and helping the client, especially if used in combination with other techniques such as reflection and attending skills.
2. Open questions invite the client to share information and talk freely.
3. Closed questions help pinpoint information, bring closure, and quickly survey a variety of content.
4. A "jackrabbit approach" to asking questions (hopping around from one topic to another) provides scattered information that is not very useful.
5. Systematic inquiry focuses on one area before the helper changes topics, resulting in more effective and efficient methods of learning about the client.

THE helper can obtain much important information from the client by asking appropriate questions. Different kinds of questions serve different purposes, and you should ask questions in a way that elicits useful information. Merely asking questions to get information is *not* the same as questioning that promotes understanding and insight by the client and helper. In this chapter, we describe how to systematically ask combinations of open and closed questions for the purpose of eliciting useful information.

Asking open questions

Open questions are an invitation for the client to talk. Operationally, an open question cannot be answered with a yes or no response. They are designed to provide the helper with information and to cause the client to think more deeply about personal behavior in the problem situation.

Open questions invite clients to talk—to open up and describe how they think, feel, and act. Many open questions begin with words such as *who, what, when, where, how,* and *why*.

> *Who* are others that are involved in this situation?
> *What* motivated you to seek help with your concern?
> *Where* were you last working?
> *When* did you first start to have these panic attacks?
> *How* long have you been having difficulties with your spouse?
> *Why* do you feel she is out to get you?
> Under *what* circumstances would you think about changing your job?

These six words are the news reporter's guide to effective interviewing for a story. For the helper, such questions increase understanding of the client's view of the world.

"Why?" is a question frequently asked in our society. At times *why* questions can help us learn more about a situation, but they can put clients on the defensive, making them feel they have to justify what has been done. Other questions are often more useful because they help the client be more concrete and specific in responding, without becoming defensive. Open questions are particularly useful when you are trying to help the client open up, talk freely about personal concerns, and provide you with specific, concrete information. The following examples start with *why* questions and then show alternatives that can provide more specific information.

- *Why question:* Why do you feel so upset?
  Alternative: What happened to make you feel upset? [Notice how this question asks the client for more specific information.]
- *Why question:* Why were you laid off from your last job?
  Alternative: What events led to your layoff?
  Alternative: How was the decision to lay you off made? [Again, notice how these open questions elicit specific information.]

Asking closed questions

Closed questions are questions that could (but don't necessarily have to) be answered with a yes or no response. Some questions begin with open words but require only one- or two-word answers. In one sense, these are also "closed" because they do *not* encourage the client to elaborate or supply more information. For example,

Who do you live with?
Where do you work?
How many brothers and sisters are in your family?

For many people, closed questions are the easiest to ask and can be extremely useful. The following are examples of closed questions for which the interviewer desires a concrete response:

Do you have a place to stay overnight where your spouse will not bother you?
Are you currently working?
Have you seen a physician or been tested for . . . ?
Are you allergic to . . . ?
Were you able to complete your work assignments satisfactorily?
Can you type?
Is there a history of [heart disease, diabetes, . . .] in your family?

Closed questions often begin with pairs of words, the second word being *you:* Do you . . . , are you . . . , were you . . . , have you . . . , should you . . . , and so forth. By not being aware of *how* you begin your questions, you may be limiting the information you get from the client.

Such questions are extremely useful and save time, particularly when all you need is a yes or no response. Notice, however, that although you *may* get a yes or no response, merely asking a closed question does not mean the client will simply answer yes or no. Highly verbal persons may give long responses to closed questions.

How you ask the question is also important in the response the client is likely to give.

- HELPER: Could you tell me how you felt in that situation?

Operationally, this question is *closed* because it could be answered yes or no. But, depending on how it is asked, such questions can be invitations for the client to share more with the helper.

Generally, be careful about two situations when asking closed questions:

1. If you start out with a series of closed questions, the client might get the idea that you will initiate the whole process and that he or she need respond with only the requested information. The client might then *expect* that you will ask all the questions and quickly learns that information is not volunteered.
2. If you have a client who is not very verbal and mainly answers yes or no to questions, you won't get much information. The client may give you only a yes or no response.

Systematic inquiry and questions

Combinations of questions asked in a systematic manner, rather than at random, often provide more useful information. If your goal is to have the client talk, ask open questions; if you want to pinpoint specific information and do not care whether the client elaborates on it, closed questions provide information quickly. In reality, a mixture of open and closed questions is best in the interview situation. *The helper must be aware of the question's probable effect on the client.* Combinations of open and closed questions are most frequently used together, depending on the helper's intent.

Systematic inquiry means seeking information in an organized manner. In a typical social interaction at an informal gathering, plentiful talk about many *different* topics occurs within a short period of time, as the verbal chatter quickly shifts from one person to another. We call this the "jackrabbit approach" because it involves hopping from one topic to another. In the helping interview, jumping from topic to topic is an inefficient way of learning about the client and gathering information.

In contrast to the jackrabbit approach, systematic inquiry is a logical, orderly method of getting information from the client. It is like a topical outline. To logically and systematically pursue a given topic, ask yourself: What kinds of information do I want to get from the client about [whatever topic or content]? The following examples show one way the helper can think about content. They also illustrate a variety of questions that could be asked in relation to what the client says.

One way to organize the information-gathering process is to inquire

Thinking questions: "What?"
These explore factual information about the client, and ideas, concepts, data, and cognitive aspects.

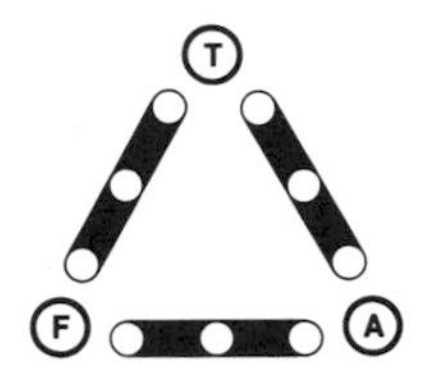

Feeling questions: "Why?"
These illuminate the client's feelings, emotions, and affective aspects.

Acting/Action questions: "How?"
These reveal things the client does, such as physicial activity, writing, reading, and talking.

FIGURE 6.1 The TFA triangle and the kinds of questions that tend to be associated with thinking ("What?"), feeling ("Why?"), and acting ("How?").

about the client's behavior (thoughts, feelings, and actions). Use the TFA triangle (Figure 6.1) as an organizer:

Thinking aspects are found by asking "what" questions to get information and data related to the client's behavior.

- CLIENT: Maybe I'm not really cut out for this job.
- HELPER: What makes you say that about . . . ?
- HELPER: What do you think you would like to do if there were no limitations?
- HELPER: What is preventing you from doing what you want to do right now?
- HELPER: What are you telling yourself about [a person or situation]?

Feelings tend to be revealed by asking *why* kinds of questions to get at emotional and affective aspects of client behavior.

- HELPER: Could you tell me more about your feelings?
- HELPER: Why do you feel you can't do [whatever] right now?

Actions become clearer through inquiry about *how* the client functions as well as activities and events one engages in.

- HELPER: How do you act when your co-workers do this to you?
- HELPER: How could you change your behavior to get a different kind of result?
- HELPER: How would you start to do [whatever] if you could begin right now?

Another way of organizing content is to inquire about specific aspects of the *situation* as it relates to the client's concern or problem. The situation is often discovered by asking a sequence of two or more questions that relate to *who, where,* and *when.*

Who questions

- HELPER: Who is involved in the situation?
- HELPER: What is it *they do* that makes you feel this way? [Such a question follows up on the first question—who is involved—by asking what they do.]

Where questions

- HELPER: Where does [this situation] happen?
- HELPER: Is it always in the same place?

When questions

- HELPER: When does [this situation] occur?
- HELPER: How often does it happen?
- HELPER: Under what circumstances does it occur?
- HELPER: Are there certain times of the day or night when it happens?

In this example, two major methods of organizing information were presented. Instead of jumping to other topics, the helper systematically inquired about important aspects of the (1) client's behavior—thoughts, feelings, and actions—and (2) the problem situation—who, where, and when events.

Systematic emphasis on specific aspects

A related way of systematically proceeding in the interview is to follow up on content from the client's last statement(s). The helper's goal is to *discover the necessary information in one area before shifting the focus of the interview to another topic*. Responses to achieve this goal might include the following:

- HELPER: Tell me more about [the topic].
- HELPER: How were you involved with [name the person or situation]?
- HELPER: What do you think about [the topic]?
- HELPER: Could you share how that made you feel?
- HELPER: What did you do about . . . ?

The helper has a tremendous influence on *what* the client talks about as well as on how it is talked about. Questions often serve as an implicit guide for the client. Depending on what the helper targets, the client quickly learns to

TABLE 6.1 How the Helper's Response Can Influence What the Client Discusses

Client Statement: "I felt awful when I found out I flunked the math test last Monday."

Content Options	Helper-Response Options
1. . . . the math test . . . (thoughts about math)	1. "Is math a unique subject for you?" "Are there other courses with which you're having difficulty?" "Was there something special about this test?"
2. "I felt awful . . . (feelings)	2. "You felt awful . . . " "Tell me about your feelings." "Awful?"
3. (No words, but implied action or inaction)	3. "How did you prepare for this test?" "What did you do prior to the test?"
4. . . . I had flunked . . .	4. "Is this the first time you have flunked?" "Have you flunked this subject before?" "How is your academic record in other courses?" "What is your previous record in math?"
5. . . . last Monday."	5. "What happened the night before the test?" "Was there something special about last Monday?"
6. (In response to the entire client statement)	6. "Um hum . . . " "Tell me more about this." Nod head as if to have the client continue talking about the concern. (*Note:* Helper-response options in item 6 allow the *client* to choose the relevant content to discuss.)

talk about these things. This is why—in some orientations to the helping process—there is a high proportion of content that relates to dreams, self-talk, early childhood experiences, parents, sexual experiences, money, jobs, interpersonal relationships, and so forth.

Beginning helpers are sometimes puzzled by what to talk about or how to proceed. Quite often, the client provides a single statement loaded with possible content that can be explored in more detail. The helper can direct the interview in radically different ways by responding selectively to the client's statements. Table 6.1 illustrates five of these ways.

As an example, the client's statement is: "I felt awful when I found out I had flunked the math test last Monday." The helper can affect the direction, or focus, of the interview by responses that lead the client to talk about (1) *thoughts* (what about math?), (2) *feelings* (how did the client feel?), (3) *actions* (how did the client study?), (4) *how* the event relates to other study activity, and (5) *when* it happened (last Monday). Yet another option (6) is present when the helper, with the use of a neutral response such as "Tell me more about it," lets the client choose the relevant content.

Chunking concepts to help the client explore areas

Beginning helpers sometimes feel at a loss about exploring ideas with clients, beyond one or two questions. The first step is to listen carefully to what the client says. After that, a concept we call "chunking" can be helpful.

Chunking is organizing key words to help explore content systematically. For example, the term "behavior" involves the concepts *thinking, feeling,* and *acting*. Thus, to explore behavior, the helper systematically inquires how the client thinks, feels, and acts in relation to the topic. Other concept chunks include the *situation* (people, data, things); *time* (past, present, future); *people* (family, friends, employer, employees); *activities* (work, leisure, volunteer); *educational experiences* (degrees, workshops, internships, conferences); *work experiences* (paid, unpaid, volunteer, formal, informal); and so forth.

Such chunks can be arranged systematically, like a matrix, to reveal connections between two or more concepts. If we combine *behavior* (thoughts, feelings, and actions) with *people* (family, friends, employer), we can help the client examine how these different groups of people think, feel, and act in regard to the topic or issue—say, changing jobs and moving to a new location. After the client responds to these questions, the helper could offer further assistance in exploring the impact of a move on *activities* (work, leisure, and volunteer), and *family* (spouse and children), and how the move would affect their *behavior* (thoughts, feelings, and actions).

When the helper *chunks* concepts into such pairs and groups them together, the available information is quickly multiplied, and topics can be examined in a useful manner. The beginning helper is encouraged to construct a matrix with two concept chunks to see how many different content items can be used in exploring a client's situation.

Summary

By asking appropriate questions in a systematic manner, the helper can assist the client in examining concerns or problems in a productive way, especially when combined with other skills such as reflection and attending.

1. Open questions
   a. Invite the client to talk and elaborate.
   b. Often begin with words such as *who, what, when, where, why,* and *how.*
   c. Cannot be answered with a yes or no response.
2. Closed questions
   a. Are helpful in pinpointing specific information in a limited amount of time.
   b. Can be answered with yes or no or very short (one- or two-word) response by the client, even though verbal clients may provide much more information.

   c. Often begin with phrases such as: Do you, are you, were you, have you, and could you, aren't you, and so forth.
   d. Limit the information that comes from relatively nonverbal persons.
3. Systematic inquiry
   a. Is a way of seeking information in an organized, outline-type manner.
   b. Involves asking combinations of open and closed questions to gather data in one area before moving on to another topic. When used most effectively, the helper relies extensively on reflecting feelings and other content, besides asking questions.
   c. Is especially useful when the helper can focus the interview on specific content such as client thoughts, feelings, and actions; the situation; or other kinds of content that are important.
   d. Chunking is a way of organizing content into a matrix so that the helper can systematically and effectively pursue important areas.

Practice

1. Find a person who agrees to help you practice asking questions and using systematic inquiry.
2. Select one situation from the list of role-plays, which follows Exercise 3. Before starting your mini-interviews, clarify specifics of the situation with your coached client: age and gender of the client, setting of the interview, and any other details essential to getting started.
3. Conduct the required mini-interviews for each exercise following these instructions. Record your interactions or have another person observe the interview, especially noting aspects related to the focus of this chapter: asking open and closed questions and systematic inquiry.
4. Following each mini-interview, ask your coached client for feedback about the interaction. In addition, after you have completed each mini-interview and discussed it with your coached client, listen to the tape or talk to your observer and answer the questions in the "Focus Checklist," which follows the exercises.
5. When you have completed Exercises 1–3, try some mini-interviews in which you practice appropriately integrating the various techniques you have learned up to now. Make your first interviews five minutes long. As you practice, extend their length to as much as ten minutes. Use the "Summary Checklist" and record all appropriate data from these more integrated interviews (Table 6.2). The "Summary Checklist" and its instructions are at the end of this chapter.

Exercise 1

Asking open questions invites the client to talk. Conduct a series of five-minute mini-interviews during which you ask only open questions. Remind your client to give fairly short answers so that *you* get the desired practice. (Have your client respond to closed questions with either a yes or no response.)

Focus Checklist on asking open questions

1. Were you able to ask only open questions?
2. What content patterns did you notice in how you asked the questions?
3. Were you able to ask several questions on a single topic before jumping to another topic?
4. When you changed topics, was there other information you could have asked the client for before changing topics?
5. Did your questions mainly relate to the client's thoughts, feelings, or actions? (If you overlooked some areas of the client's thoughts, feelings, or actions, practice asking questions that elicit client response to areas you omitted.)

Exercise 2

Asking closed questions produces yes or no or very short (one- or two-word) responses. Conduct a series of two- or three-minute mini-interviews and ask *only* closed questions. Ask your coached client to give only yes or no responses.

Focus Checklist on asking closed questions

1. Were you able to ask only questions that could be answered with a yes or no response from the client?
2. What did it feel like when the client only gave a one-word or very short answer?

For more practice in asking open questions, write each closed question and then write two or three open alternatives.

Exercise 3

Systematic inquiry searches for specific content. Conduct a series of five-minute mini-interviews in which you practice asking combinations of open and closed questions to systematically pursue specific content. Remind your client to give relatively short answers so that *you* get the practice. For added awareness, tell your coached client to give *only* a yes or no response if you ask questions that *could* be answered with a yes or no. This will help you think about what you are doing. Make a chart similar to the one in Table 6.2 and record your open and closed questions. Writing key words will help you see how well you are following the client's content or whether you tend to jump from topic to topic.

TABLE 6.2 Recording Key Words and Phrases from the Helper and Client

Key Words and Phrases from the Helper (H) and Client (C)	**Emphasis on T**	**F**	**A**	**Open Question**	**Closed Question**	**Systematic Inquiry**	**Changed Content**
C: I was shocked when I found out.		✓					
H: How did you hear about this?	✓			✓			
C: Well, Jamie is my best friend, and she let it slip.	✓						
H: Were you able to talk to your parents about it?			✓		✓	✓	
C: Uh . . . yeh . . . kinda [looking toward the floor].	✓	✓					
H: What was their reaction?	✓			✓		✓	
C: They couldn't believe I had done it.	✓						
H: How did you feel right then?		✓		✓		✓	
C: I just wanted to crawl in a hole and hide.		✓	✓				
H: What do you want to do about it at this point?			✓	✓			✓

Focus checklist

1. How many closed questions were in your interview sequence? How many open? (If you asked many closed questions, you can profit by more practice in asking only open questions.)
2. Did you ask several questions on a single topic before switching to another?

3. Did you jackrabbit and switch topics after only one or two questions? If you did, practice asking three or four questions that are related to that topic. If you are doing this with a fellow student, try a couple minutes of brainstorming on each topic area to see how many more questions you could ask on each topic.
4. What kinds of organizers or chunking matrix did you use in your interview—for example, inquiring into the client's thoughts, feelings, and actions; discovering more about the situation (who, what, where, and when); following major elements in the client's words; or other organizers?
5. Were there *patterns,* or sequences, of questions you noticed in your interview?

Role play

1. Your client lacks motivation in school and has suggested that not all is well at home, especially when the stepfather comes in late at night.
2. The client has been coming in late and missing work on an increasingly frequent basis. Reasons for tardiness and absenteeism are not clear.
3. The client thinks she is pregnant and is upset. She has no idea what to do and cannot think straight. Or, your client's girlfriend is pregnant, and he is distraught and sees all his plans going astray.
4. Your client (age thirty-five to fifty-five) is dissatisfied with life. Present circumstances are not very rewarding, and your client is beginning to search for "something more meaningful to do."
5. The client, who has been feeling rather sluggish and not up to par, has decided to go to the doctor but is afraid of what the results of the checkup might be.

Summary checklist

Beginning in this chapter and continuing through Chapter 21 are Behavior Summary Checklists to help you review and *appropriately* integrate skills learned up to that point in each chapter. Use the role plays (or create your own), conduct a five-minute mini-interview, and assess it using the skills and the Behavior Summary Checklist.

How many of the following elements did you use, and in what ways did you use them during your more integrated mini-interviews?

1. Structuring the helping relationship
2. Focusing on thoughts, feelings, actions, or TFA combinations
3. Listening for understanding and nonverbal aspects
4. Asking open questions
5. Asking closed questions
6. Systematic inquiry

Summary Behavior Checklist Key words and phrases of the Client (CL) and the Helper (H) T F A	
Focus or Emphasis — T	
Focus or Emphasis — F	
Focus or Emphasis — A	
Structure	
Nonverbal aspect	
Open question	
Closed question	
Systematic inquiry	
Clarification	
Reflection	
Confrontation	
Use of silence	
Client's concern	
Goals	
Procedures	
Evaluation	
Termination	
Followup	
Strategies	

Instructions

To assess your interview skills, make a worksheet similar to the Behavior Summary Checklist. In the *left column* on the checklist, make a transcript of key words and phrases or complete sentences used by you and your client during the mini-interview. Then, *for each* transcript response, put a check in columns that are relevant. You might check from one to several items for each response. Ignore columns with content area that has not been discussed.

When you have completed the Behavior Summary Checklist, review it carefully. Look for patterns where you emphasized or ignored content. Then write a brief summary of the mini-interview outlining your strengths, limitations, and suggestions for improving skills.

References and recommended reading

Evans, D. R., Hearn, M. T., Ulhemann, M. R., & Ivey, A. E. (1989). *Essential interviewing* (3rd ed.). Pacific Grove, CA: Brooks/Cole. This superior self-instruction book contains sample client content followed by options that could be used. Each option is explained in terms of appropriate use. We highly recommend this book for those who would like more practice on helping skills similar to those in Part 2 of this book.

Hackney, H., & Cormier, L. S. (1988). *Counseling strategies and interventions* (3rd ed.). Englewood Cliffs, NJ: Prentice-Hall.

Ivey, A. E. (1983). *Intentional interviewing and counseling*. Pacific Grove, CA: Brooks/Cole.

• Reflection and Clarification

7

Key points

1. Reflection is a way of mirroring the essence of the client's communication, as well as putting into words what the client cannot say.
2. Reflection can focus primarily on the client's thoughts, feelings, actions, or combinations of these components of behavior.
3. Feelings are not necessarily more important than what a person thinks or does.
4. Reflection often helps the client see how thoughts, feelings, and actions combine to suggest a direction and purpose to one's behavior. This awareness is one of the first steps in changing behavior.
5. Clarification helps reduce vague or uncertain aspects of the client's behavior and the situation.
6. Clarification of the client's meaning may occur through reflection, restatement, paraphrase, and questions that are linked to the client's nonverbal behavior.
7. The helper influences content of the interview by selective attending and responding to specific aspects of the client's communication.

WHEN the helper reflects the essence of the client's communication in a systematic way, this helps clarify concerns and specific aspects of the client's behavior in the specific situation.

Reflection of thoughts, feelings, and actions

Some people have the mistaken impression that the helping process deals *only* with people's feelings. Although feelings are important, they are not more important than thoughts and actions. The effective helper, recognizing the importance of all dimensions of behavior, deliberately focuses on specific aspects of behavior that are critical to understanding the client's situation.

Reflection is a way of putting into words both what is said and what the client cannot say. It includes verbal and nonverbal aspects of communication—*what the client says* (the actual words), *how the client says things* (e.g., nonverbal or subvocal aspects of communication, including tempo of speech, pitch of voice, and hesitations), and the *physical or bodily aspects of communication* (e.g., gestures, eye and facial expressions, body posture, and movement).

Good reflection clarifies meaning

Reflecting does *not* mean that the helper parrots what the client says:

- CLIENT: I'm feeling really down today.
- HELPER: You're feeling really down today.

This is simply a *restatement* of what the client said and adds nothing to what the client communicated verbally. Reflection is a way of summarizing parts of the client's behavior—of mirroring the essence of the client's thoughts, feelings, and actions. This usually provides a clearer picture for both client and helper. The helper reflects important aspects of the client's behavior to the client.

When using reflection, the helper does two things: (1) expresses the core or essence of the client's behavior (2) in a few words. The following is an example of an effective use of this technique:

- CLIENT: I really like working with people. I especially enjoy my teaching experiences because I can have a real impact on school-age children when they are so impressionable. It's a great feeling to help a child turn around and do something positive. But with buying a house and trying to provide for my family and keep up with inflation, I could earn more and work less as a grocery store cashier . . . and we really need the money to survive.
- HELPER: You're really torn between wanting to teach and needing more money.
- CLIENT: Yes! That's it exactly!

Notice two qualities in the above reflection. First, the helper identified the client's *feeling* ("You're really *torn*") about content (the two competing options of teaching versus earning more money). Second, the helper's *short* reflection helped clarify the client's central concern. This allowed the client to see the situation more clearly than before. Seeing the problem clearly is the first and most important step in the behavior-change process.

The helper's ability to put concerns in a clean, clear, concise summary is especially helpful because it clarifies the picture for the client. *Thoughts* can be summarized and ordered logically, expressed and unexpressed *feelings* are clearly identified, and the client's *actions*—or lack thereof—are seen more clearly than before. The critical threads and themes that run through the client's behavior are often seen for the first time. When the helper clearly reflects the essence of the client's behavior, the client often gains *insight* because matters are seen more clearly than before. Critical connections between ideas, emotions, and events are identified. Sometimes, clients can develop a plan for handling problems that seemed impossible before this insight occurred, which helps explain why person-centered or insight-oriented approaches to helping work for these individuals.

Reflecting client thoughts, feelings, and actions

Reflection is a powerful indicator to the client that the helper is really listening and attending carefully to all possible stimuli. Three broad questions can serve as guides for the helper in trying to understand how the client perceives concerns:

1. What does the client *think* about the concerns?
2. What does the client *feel* about these concerns?
3. What does the client *do* (or how does the client act or not act) about these concerns?

The following situation illustrates the range of responses the helper can make to what the client says:

- CLIENT: The project seems so tough I don't see how I can possibly complete it on time. If I don't get it done, the boss will really get on my case. I just don't know what I'm going to do.

Before proceeding, think about how *you* might respond to this client if you were trying to focus on each of the following. Use combinations of reflection and questions as you write your responses under the following headings on a separate sheet of paper:

1. Client thoughts: ______________________________________

   ____

2. Client feelings: ______________________________________

   ____

3. Client action: ______________________________________

   ____

4. Combinations of the client's thoughts, feelings, and actions:

   ____

   ____

The following are examples of some possible helper responses that specifically focus on client thoughts, feelings, and actions. See how these compare with the responses you just wrote.

Focus on client thoughts or information

- HELPER: You're expecting the boss won't be understanding, and you see no way to avoid this.
- HELPER: What about the project seems so tough to you?
- HELPER: When does the project need to be completed?

Focus on client feelings

- HELPER: You're *stuck* in your tracks and don't know where to go.
- HELPER: You're at your wits' end. It's a *frightening* prospect to you.
- HELPER: The situation seems pretty *hopeless*.

Focus on client actions

- HELPER: At this point, you don't know how to proceed.
- HELPER: What have you done about the situation?
- HELPER: You're *caught* between the project deadline and not knowing what to do.
- HELPER: How could you handle this problem better?

Focus on combinations of the client's thoughts, feelings, and actions

- HELPER: Being criticized by your boss and the threat of losing your job seem pretty scary to you. [This reflects the client's thoughts and feelings about possible actions.]
- HELPER: The fear of what might happen prevents you from doing anything about it. [This reflects both feelings and inaction of the client.]
- HELPER: You feel overwhelmed . . . [pause]. What do you think you could do to start to work on the project? [The helper acknowledges the client's feelings and then asks the client to begin thinking about possible actions.]

These examples illustrate the helper's options for focusing the inquiry or reflection on the client's thoughts, feelings, and actions. To a great extent, the helper's focus influences the direction of the interview.

In certain parts of the interview, the helper might choose to focus on one particular aspect of behavior: to see what the client is *thinking,* to explore *feelings* and emotions in some depth, or to examine *action* options about how to proceed. Some clients, as well as helpers, tend to be mostly cognitive, affective, or action-oriented (Hutchins, 1984). As illustrated in Chapter 3, an awareness of yourself in a helping role, in addition to behavior of the client, can pay big dividends. For example, ask yourself these kinds of questions as you practice and listen to various interviews:

1. Do I tend to focus on or ignore the kinds of thinking in which the client engages?
2. Do I tend to emphasize or ignore the client's feelings and emotions?
3. Do I tend to inquire about or ignore the client's actions or potential actions, or inaction?
4. Is the interview balanced in terms of a cognitive, affective, and action orientation, or is it somewhat lopsided with an emphasis on one or two aspects while ignoring others?

The alert, self-aware helper can detect these orientations by paying close attention to what the client says, how the client says it, and what the client does. Also, when you draw the client's TFA triad (Chapter 3), you'll be likely to see the client's orientation reflected in the triad pattern.

1. *What the client says:* Does the client talk about ideas and concepts, analyze things, discuss theories, rationalize behavior, and the like? If so, these indicators suggest that the client has a thinking orientation. The client who talks about feelings, affect, and emotions may have a feeling orientation. And the client who talks about activities or things that have been done may have an action orientation.

2. *How the client says things:* Is the client, for example, enthusiastic, interested, excited, calm, depressed, anxious, concerned, frightened, uncertain, or hesitant? Observing verbal and nonverbal behavior (facial and body expressions) related to how things are said or not said can provide important information about the client's feelings.

3. *What the client does:* Does the client take action and get things done? Procrastinate? Act in an overly cautious or hesitant manner? Jump right into things? Does the client do things alone? Is the client involved in few activities, a narrow range of activities, or virtually everything that comes along? Such observations provide information about the client's habits, activity level, energy, and involvement with ideas, people, and things.

Sometimes, in a good reflection of feeling, the helper may discern the client's orientation almost exclusively from nonverbal cues:

- CLIENT: [looking toward the floor, somewhat slumped in body position, not looking at the helper, and perhaps shuffling feet, wringing hands, or tapping fingers on the chair].
- HELPER: You seem to be [upset, nervous, anxious]. Could you share your feelings with me . . . [pause]?

This may also be an appropriate time to *structure* aspects of the helping process that relate to confidentiality if you think the client cannot decide on sharing very personal and sensitive information (see Chapter 4 on structuring):

- CLIENT: [same situation as above, where you can almost see the client struggling with whether to talk about very sensitive issues].
- HELPER: You seem to be struggling with something really tough to talk about . . . [pause]. Remember that what we talk about here is confidential.

The helper may then go on to clarify any limits of confidentiality that relate to the specific setting or organization, if these have not been discussed previously.

Of course, you must be aware of social, educational, and cultural differences that may make a difference in how you interpret the client's behavior (Pederson, 1985; Sue, Akutsu & Higashi, 1985). By being alert for cues to the client's behavioral orientation, the helper may want to adjust the interaction to meet the client's needs more adequately. If the client cannot talk about feelings, approach the client in more cognitive ways first, getting into more affective areas after a greater degree of trust has been established. Similarly, if the client appears to be more action-oriented, approach topics through inquiry about activities and use this information as a base from which to gain data about feelings.

There is no one right way of reflecting. The helper must be aware of options and use them to respond to important aspects of the client's concerns. Also, there may be times when the helper wishes to systematically pursue thoughts, feelings, or actions in some detail, and this is fine. However, exploring specific areas of content more completely, rather than jumping from one area to another, is more useful. For example, exploring client feelings and emotions in some depth on a given topic may be desirable before moving on to another area (review the discussion in Chapter 6 on systematic inquiry). By deliberately and genuinely adapting to the client, the helper is likely to increase the client's perception of the helper as one who really cares. The belief that another important person is interested and really cares can be powerful medicine in influencing the client's own personal resolve to overcome major problems, including physical health as Cousins (1979) suggests.

Clarifying with questions

As mentioned in the discussion of questions and systematic inquiry (Chapter 6), the helper can clarify much of the client's meaning by asking questions that start with the words *who, what, when, where, why,* and *how.* Furthermore, the helper can pinpoint specific information about the client by asking closed questions, which are answered yes or no.

The effective helper can use questions to clarify the client's meaning in several ways. By questioning generalities, the helper gains an understanding of what the client means. Consider the following:

- CLIENT: The work I do is a real bummer.

Probably many of us have used similar words to describe a difficult or awkward task. But we really do not know what *this* particular person means by the phrase *real bummer,* and so the effective helper will use questions, such as the following, to learn exactly what the client means:

- HELPER: Real bummer? [The helper restates key words used by the client.]
- HELPER: What do you mean by real bummer? [The helper asks a full-sentence question, again using the client's key words.]
- HELPER: What about your work is a bummer? [The helper's intent is to find out details about the work that the client considers to be a bummer.]
- HELPER: You don't get much satisfaction from the work you do. [The helper interprets.]

In the last example, the helper interprets the meaning behind the client's actual words. There are two likely client responses to this. If the helper's interpretation is correct, the client might say, "Yes, that's right" or "That's for sure"; or if it is incorrect, the client will likely clarify or enlarge upon what was meant. *Clarification is what the helper does to find out what the client means by what is said or not said,* often picked up by careful observation of the client's nonverbal behavior.

Summary

1. Reflection mirrors important aspects of the client's thoughts, feelings, and actions.
2. Good reflection usually includes both verbal and nonverbal aspects of the client's communication.
3. Reflection clarifies content because it expresses the client's communication more concisely than before, frequently helping the client to see matters more clearly than before (insight).
4. By systematically attending to certain aspects in the interview, the helper influences both what is talked about and how the client talks about it.
5. In assisting the client, the helper must be aware of individual, social, and cultural behavior patterns that affect the client's behavior in the specified situation.
6. There are many ways of clarifying aspects of the client's behavior, including reflecting verbal and nonverbal aspects, restating key words and phrases, paraphrasing, asking open and closed questions, and using systematic inquiry.

Practice

1. Find a person who agrees to help you practice using reflection. Assume that you and your coached client have been talking for about ten minutes.
2. Select one situation from the list of role plays, which follows these instructions. Before starting your mini-interviews, clarify specifics of the situation: age and gender of the client, setting of the interview, and any other details essential to get started. For this activity, *ask your client to express a lot of feeling* but not to make long speeches during the role-playing sessions. (You're the one who needs the practice.)
3. Conduct a five-minute mini-interview for each role-playing situation on the list. Record your interaction or have another person observe the interview, especially noting aspects related to reflection.
4. Following each mini-interview, ask your coached client for *feedback* about the interaction (see Chapter 3 on feedback). Then listen to the tape or talk to your observer and answer the questions in the "Focus Checklist," which follows the list of role-playing situations.

Role play

1. Your client has just broken up with a lover or partner. They have had a fairly good relationship for several months until last week.
2. Your client, who has recently suffered the death of a spouse, comes to you fairly upset and not knowing what to do. Lacking a formal education and facing the possibility of having to live on the limited money from social welfare or the government, this individual is wondering about moving to another part of the country to be closer to other family members.
3. Your client, a student who has just completed the freshman year, received D's in three science courses and realizes it will be impossible to get into medical school with such grades. The client is distressed and says the pressure at home is enormous because both parents are physicians and have been counting on their child to carry on the medical tradition in the family.
4. The client, a thirty nine-year-old male, is feeling tremendous pressure to spend more time with his wife, who says they are becoming strangers. He works seven days a week, including night hours, to support the family. He expresses that he truly feels as if he's caught in a vise.[1]
5. Although your client knew that the parents had not been getting along well for some time, the recent news of their decision to divorce was a shock. The client cannot think about anything else and hasn't slept for the last two nights.
6. Your client, a school-age child, was referred to your office for fighting a friend in the hall. Entering your office, the child appears to be distressed and angry.
7. The industry in which your client has been working for the last six years is closing down. The client is anxious about not finding work in the area—especially a well-paying job.

Focus checklist

1. How may times did you use reflection in the mini-interview?
2. Was the focus of the reflection primarily on the client's thoughts, feelings, and actions or on other aspects of behavior?
3. Were there *patterns* as you reflected thinking, feeling, and acting content from the client?
4. Did you help the client systematically explore and clarify one topic area before moving to another topic? Or did you tend to jump from one area to another?
5. Did you offer your clarifications and reflections in a caring and concerned manner?

[1] Thanks to Linda Dudley.

6. Did your reflections mirror the essence of what the client was expressing?
7. Did you reflect the client's behavior *concisely,* or did you tend to be wordy, rambling, or lecturing?
8. What other ways did you use to clarify the client's meaning (restatement, paraphrase, questions, etc.)?

Summary checklist

After answering the questions in the "Focus Checklist," try some mini-interviews in which you practice appropriately integrating the various techniques learned up to now. Use the following checklist and record all appropriate data from these more integrated interviews.

How many of the following elements did you use, and in what ways did you use them during your more integrated mini-interviews?

1. Structuring the helping relationship
2. Focusing on thoughts, feelings, actions, or TFA combinations
3. Listening for understanding and nonverbal aspects
4. Asking open questions
5. Asking closed questions
6. Systematic inquiry
7. Clarification
8. Reflection

Instructions

To assess your interview skills, make a worksheet similar to the Behavior Summary Checklist. In the *left column* on the checklist, make a transcript of key words and phrases or complete sentences used by you and your client during the mini-interview. Then, *for each* transcript response, put a check in columns that are relevant. You might check from one to several items for each response. Ignore columns with content area that has not been discussed.

When you have completed the Behavior Summary Checklist, review it carefully. Look for patterns where you emphasized or ignored content. Then write a brief summary of the mini-interview outlining your strengths, limitations, and suggestions for improving skills.

References and recommended reading

Benjamin, A. (1981). *The helping interview* (3rd ed.). Boston: Houghton Mifflin.

Cousins, N. (1979). *Anatomy of an illness as perceived by the patient.* New York: Norton.

Egan, G. (1990). *The skilled helper* (4th ed.). Pacific Grove, CA: Brooks/Cole.

Evans, D. R., Hearn, M. T., Uhlemann, M. R., & Ivey, A. E. (1989). *Essential interviewing* (3rd ed.). Pacific Grove, CA: Brooks/Cole. This superior self-instruction book contains sample client content followed by options that could be used.

Summary Behavior Checklist

Key words and phrases of the Client (CL) and the Helper (H)

T F A

Focus or Emphasis	
T	
F	
A	

Structure	
Nonverbal aspect	
Open question	
Closed question	
Systematic inquiry	
Clarification	
Reflection	
Confrontation	
Use of silence	
Client's concern	
Goals	
Procedures	
Evaluation	
Termination	
Followup	
Strategies	

Each option is explained in terms of appropriate use. We highly recommend this book for those who would like more practice in helping skills.

Gaylin, W. (1976). *Caring*. New York: Knopf. Both *Caring* and *Feelings* (see below) are fascinating explorations into the roles of caring and feeling in our society.

———. (1979). *Feelings: Our vital signs*. New York: Ballantine Books.

Hutchins, D. E. (1984). Improving the counseling relationship. *Personal Guidance Journal, 63*, 572–575.

Pedersen, P. (Ed.). (1985). *Handbook of cross-cultural counseling and therapy.* Westport, CT: Greenwood Press.

Rogers, C. R. (1961). *On becoming a person.* Boston: Houghton Mifflin. Rogers was the founder and developer of the client-centered approach to counseling and psychotherapy. Earlier methods are called nondirective, whereas later methods are called person-centered.

———. (1986). Reflections of feelings. *Person-Centered Review, 2*, 375–377.

Sue, D., Akutsu, P. D., & Higashi, C. (1985). Training issues in conducting therapy with ethnic-minority-group clients. In P. Pedersen (Ed.), *Handbook of cross-cultural counseling and therapy* (pp. 275–280). Westport, CT: Greenwood Press.

The Effective Use of Silence

8

Key points

1. The helping interview is one place the client can think quietly, in a relaxed manner, about all aspects of the situation.
2. During certain periods of silence, the client can gain tremendous insight and growth. Thoughts and feelings that have never before been dealt with often arise, and previous thoughts and feelings can be sorted out.
3. The helper should allow the client's "wheels of thought and feeling" to turn without interruption.
4. If appropriate periods of silence occur, the client will often add important content without the helper having to ask follow-up questions.
5. Becoming comfortable with appropriate periods of silence provides freedom that allows the helper to attend to all aspects of the client's behavior.

FOR a beginning helper, one of the most difficult tasks to learn is *avoiding talking and letting silence occur at appropriate times during the interview*. Many people are so used to expressing their own opinions, giving advice, or telling personal experiences that they tend to carry over this behavior into the interview. Furthermore, many people become uncomfortable if more than three to five seconds of silence occur while talking with others.

Why use of silence is important

For beginning helpers, fear of silence can result in the following types of behaviors:

- Not listening fully to the client
- Ignoring important relationships among client thoughts, feelings, and actions
- Interrupting before the client is finished talking or thinking things through
- Randomly sputtering out any thought to fill the silence
- Cutting off productive client thought about significant content that has just occurred in the interview (insight)
- Mentally rehearsing thoughts that might be said, while the client is still talking (thus not giving full attention to the client)

Any of these events may happen if the helper has not learned to be comfortable with letting silence occur at appropriate times during the helping interview.

How silence helps the helper

Becoming comfortable with silence and allowing silence to occur benefit the helper in these ways: You can

- Listen more effectively.
- Pay better attention to the client's nonverbal behavior.
- Better assess relationships among the client's thoughts, feelings, and actions.
- Follow more thoroughly what the client says.
- Ask more insightful questions.
- Reflect content in a more comprehensive and helpful manner.
- Do a better job of focusing attention on the client's most critical concerns.

With appropriate use of silence, the helper's responses and comments are much more likely to help the client deal with critical variables that arise during the process of change. Such variables might include the client's special meanings of words and ideas, how things relate to each other, priorities, values, content brought to awareness for the first time, and a host of other concerns.

When to use silence in the interview

A general guideline is, *Don't rush the client into talking*. Let the client set the pace of the interview, especially during the initial stages when trust and confidence are being built. The helping interview may be one of the few opportunities the client has to fully express thoughts and feelings without being rushed or pressed to perform. This luxury of unhurried time allows more complete expression than is typical in most clients' interactions. Thus, within a good helping relationship, the interview will proceed at a pace with which the client is comfortable.

At times, there can be awkward periods of silence, often resulting in the client saying something to fill in the void. If the helper can model an unrushed silence, the client tends to become more comfortable with it, thereby enabling better thinking through and reflecting on important concerns.

When silence is being brought into play, the helper can use a number of cues to relate silently to the client. The helper can

Nod the head casually, indicating approval of what the client is doing.

Vocalize sounds (such as uh-huh, yes, and m-hmm) to indicate support for what the client is doing now and *to encourage the client to continue along the same line.*

Maintain eye contact with the client (sometimes coupled with a nod of the head or some vocalized sound), indicating further that what has been said is in the right direction and that what is happening right now (silence) is fine.

Thus, combinations of eye contact, vocal, and other nonverbal expressions are ways of letting clients know that you are with them and following what has been said. Such signals let clients feel free to continue with their thoughts and feelings, secure in the knowledge that you are devoting full attention to the topic. Often, just allowing a few moments (five to eleven seconds) of silence to follow a client statement will encourage continued talking and disclosure of additional relevant content. If helper and client are in a good relationship, silence—by itself—can be a strong motivator to add more content to what has been said; frequently, these clients will continue to add information relevant to the topic if the helper merely allows silence (and thought) to occur and does not rush the interview.

Critical times: "The wheels are turning"

Allowing silent time to occur is especially important when the client or helper has said something of critical importance. This might include the client arriving at a moment of insight such as an awareness of the relationship between events or seeing the relationship among thoughts, feelings, and actions for the first time. At times such as this, when important thinking occurs, the most important technique that can be used is silence. *Let the client think about the situation without interruption!* At such times, you can virtually see the "wheels turning" in the brain. Let it happen! Chances are good that more important things are happening during this silence than any words could express. So use the silence and allow this special process to be completed, at which time the client will usually voluntarily relate the unique thoughts and feelings that have just occurred.

When the process has been completed, the helper can inquire about what has been going on with the client. The following kinds of responses might prove useful:

- HELPER: You seemed to be in pretty important or meaningful thought.
- HELPER: What's been happening during these last few moments?
- HELPER [in an inquiring tone of voice]: You seemed to be thinking [or feeling] something pretty deeply.
- HELPER: Your silence indicated to me that something significant happened.

In general, long periods of silence (more than a few minutes) are not considered useful in the helping interview. To assist the client, the helper must be aware of the client's thoughts and feelings, and this cannot occur satisfactorily unless the client shares them. However, for an interesting and different point of view, read Calia and Corsini's *Critical Incidents in School Counseling* (1973). One of the situations discussed therein illustrates a case in which practically the entire interview consisted of silence between the helper and client.

Summary

1. It is important to learn the art of nontalking and let silence occur at appropriate times during the interview.

2. The helper's failure to use silence effectively leads to interruptions while the client is speaking, the cutting off of productive client thought, or rushing the client. In addition, if the helper rehearses what might be said, rather than paying attention to the client, important client content is likely to be missed.
3. Appropriate use of silence allows the helper to be more effective. Freedom from the pressure of having to fill silent moments with words results in more effective listening, reflecting relevant content, more accurate understanding of content, asking more effective and systematic questions, and focusing attention on more critical concerns of the client.
4. The helping process may be the client's only opportunity to think through ideas, express feelings, and problem-solve in an unhurried manner.
5. For some clients, silence can be a powerful incentive to add content to what has already been said in the interview.
6. Often, following client insight, short periods of silence can help the person think through and integrate meaningful material.

Practice

The goals of these activities are fourfold:

1. To reduce your discomfort and nervousness as a helper when silence occurs during the interview
2. To increase your comfort as a helper in allowing productive periods of silence to occur in the interview
3. To teach you to recognize when silence is appropriate
4. To help you to use silence effectively in focusing on behavior of the *client* rather than on yourself

Instructions

Select a partner who agrees to help you practice using silence. Assume that you have already conceptualized the concerns of the client in a previous session. Explain to your partner that *both* of you will probably feel very awkward because what you will be doing will be exaggerating periods of silence.

The goals of Exercises 1 and 2 are to reduce your discomfort or nervousness and increase your comfort in letting periods of silence occur. As a result, you will be able to recognize when silence is appropriate and use it to focus on the client's behavior. If you do these exercises correctly, *your interviews will seem awkward* and not like good interviews at all. This is just the way it should be. To practice the effective use of silence, do *not* try to have an appropriate interview when doing these exercises.

Exercise 1

Select a role-playing situation from the list of role-playing situations that follows these instructions, and conduct a three-minute mini-interview with your coached client. Record your interaction. During your role playing, whenever the coached client finishes saying something, don't say anything. Allow five seconds of silence *after every client statement* before you say anything. Encourage your coached client to keep statements short so that you can practice using silence. When you have completed three minutes of interview, play back the tape and listen to the interaction. If it seems unnatural and uncomfortable to you both, you have most likely done the exercise correctly.

Continue practicing this activity until you become comfortable letting five-second periods of silence occur. As you review your tape, pay special attention to your response. In the "Focus Checklist," note the kinds of responses.

Exercise 2

Proceed as you did in Exercise 1, but this time allow ten, fifteen, and twenty seconds of silence to occur after the client completes a thought. This may be difficult but try it for several five-minute mini-interviews. This kind of *extreme practice* will make you feel more comfortable with shorter periods of silence, allowing you to do a better job of listening to the client, reflecting relevant content, asking relevant and systematic questions, and so on.

As in the past, record each mini-interview. When you play it back, pay particular attention to your responses to the client's communication. As a beginning helper, you should become more and more effective in picking up on specific major verbal and nonverbal content from the client's communication. In the "Focus Checklist," make a note of responses you could have made that would improve your attending and following behavior.

Role play

1. This client has been having a relationship with a person of the opposite sex. After three months of going together steadily, the client proposed marriage, only to be told that marriage was out of the question. The client, stunned by the response, seeks your help to sort feelings out.
2. A sixteen-year-old client reports to you that family relationships are not good. As the helper, you explore the difficulties involved in this situation.
3. A thirty-five-year-old client who has been out of work for five months expresses concern about being unemployed and the hardship that this brings in terms of feelings of low self-worth, family problems, and fears of a bleak future.

4. This client, whose seven-month marriage has, until now, been a happy one, is concerned by recent lack of spousal attention.
5. You are seeing a child because of repeated problems (verbal abuse and fist fights) with others. These fights have occurred with increasing frequency. You have heard that the child's parents are getting a divorce.

Focus checklist

1. Were you able to allow the desired five, ten, fifteen, or twenty seconds of silence to occur following client statements?
2. What happened with the client when you allowed longer periods of silence to occur? (What did the client say? How did the client feel?)
3. How did you feel during these pauses?
4. What were you thinking during the pauses?
5. Did you use the periods of silence to mentally rehearse (practice) what you would be saying next, or did you focus on the client?
6. Could you use the silence to reflect on the client's nonverbal behavior as well as what was said and how it was said?
7. As you became more comfortable allowing longer periods of silence to occur, were you able to respond more appropriately to both verbal and nonverbal aspects of the client's behavior?
8. Did your questions relate to the client's communication or did you say the first thing that occurred to you?
9. If short pauses (five to ten seconds) made you uncomfortable, did you practice even longer periods of silence (fifteen to twenty seconds) to help you (eventually) overcome your feelings of discomfort?

Summary checklist

After you have become relatively comfortable letting ten-second periods of silence occur, add all other relevant skills to the mini-interviews. As you become more comfortable with silence, such skills will come much more easily because of the reduction of pressure. You will be able to listen to what the client says more effectively, observe more of the client's nonverbal behavior, note more of the thinking–feeling–acting relationships, and so forth. In short, with your increasing comfort in using silence, you will also continue to make increasingly effective verbal responses in the helping interview. When you practice integrating other skills, your goal shifts from *artificial,* intensive practice of one or two techniques to *appropriate integration* of all techniques learned previously. After taping your interview, use the Behavior Summary Checklist to record all appropriate data from these integrated interviews. Besides recording key words and phrases, each time the client finishes speaking, record the number of seconds you allowed to elapse before you spoke your next words.

As you do the more integrated interviews, note how many of the following elements you used and in what ways you used them.

1. Structuring the helping relationship
2. Focus on thoughts, feelings, actions, or TFA combinations
3. Listening for understanding and nonverbal aspects
4. Asking open questions
5. Asking closed questions
6. Systematic inquiry
7. Clarification
8. Reflection
9. Use of silence

Instructions

To assess your interview skills, make a worksheet similar to the Behavior Summary Checklist. In the *left column* on the checklist, make a transcript of key words and phrases or complete sentences used by you and your client during the mini-interview. Then, *for each* transcript response, put a check in columns that are relevant. You might check from one to several items for each response. Ignore columns with content area that has not been discussed.

When you have completed the Behavior Summary Checklist, review it carefully. Look for patterns where you emphasized or ignored content. Then write a brief summary of the mini-interview outlining your strengths, limitations, and suggestions for improving skills.

References and recommended reading

Calia, V. F., & Corsini, R. J. (Eds.). (1973). *Critical incidents in school counseling*. Englewood Cliffs, NJ: Prentice-Hall.

Egan, G. (1990). *The skilled helper* (4th ed.). Pacific Grove, CA: Brooks/Cole.

Gaylin, W. (1976). *Caring*. New York: Knopf.

———. (1979). *Feelings: Our vital signs*. New York: Ballantine Books.

Rogers, C. R. (1961). *On becoming a person*. Boston: Houghton Mifflin. Rogers is the founder and developer of the client-centered approach to counseling and psychotherapy.

———. (1975). Empathic: An unappreciated way of being. *Counseling Psychologist*, *5*, 2–10.

———. (1986) Reflections of feelings. *Person-Centered Review*, *2*, 375–377.

Zimbardo, P. G. (1977). *Shyness*. Reading, MA: Addison-Wesley.

Summary Behavior Checklist

Key words and phrases of the Client (CL) and the Helper (H)

T F A

Focus or Emphasis: T	
Focus or Emphasis: F	
Focus or Emphasis: A	
Structure	
Nonverbal aspect	
Open question	
Closed question	
Systematic inquiry	
Clarification	
Reflection	
Confrontation	
Use of silence	
Client's concern	
Goals	
Procedures	
Evaluation	
Termination	
Followup	
Strategies	

Confrontation

9

Key points

1. Confrontation challenges the client to face reality and consider alternatives.
2. Confrontation can be positive and helpful when done in a caring way.
3. Confronting a client with discrepancies in thoughts, feelings, and actions often leads to greater awareness of the reality of particular behavior.
4. If the client is to take positive action, discrepancies between one's current and desired (goal) behavior must be confronted and resolved.
5. The client quadrant is useful in examining the client's current behavior and consequences as well as exploring alternatives.

> CONFRONTATION is a direct technique in which the counselor challenges clients to face themselves realistically. . . . Confrontation is an open, honest identification of the client's self-defeating patterns or manipulations. . . . The counselor shares how those inappropriate behaviors produce negative consequences in interpersonal relationships. It is a challenge for the client to integrate the conflicting aspect of his or her being. (Leaman, 1978, p. 630)

Confrontation, caring, and feedback

Although some people have the idea that confrontation means a direct "hit below the belt," confrontation can be one of the most helpful techniques available for the client—if used correctly. Helpful confrontation has two major characteristics:

1. *Caring and concern*: The helper must confront the client in the spirit of caring and concern for the welfare of the client as an individual. The helper reflects this caring (or empathic concern) in such a way so that the client does not misunderstand. The client must clearly perceive that the helper cares enough to challenge and confront behavior directly and honestly. Furthermore, the helper will assist the client in understanding and working through the behavior to successful resolution.

2. *Specific, descriptive, well-timed feedback*: Helpful confrontation provides the client with objective feedback about important aspects of behavior. This feedback is specific and descriptive rather than vague. It is well timed and takes into consideration the client and personal circumstances. The helper identifies the behavior and brings it to the client's awareness (see Chapter 3 on feedback). Because the helper confronts the client in a caring manner and describes rather than evaluates the behavior, the client is much more likely to hear and accept what the helper says.

- POOR: You're a terrible listener.

This is a poor confrontation because it only vaguely describes the client's behavior and is emotionally loaded with an evaluative opinion, *terrible listener.* The chances are good that this kind of behavior would produce a negative or defensive emotional response from the client.

- BETTER: When I asked you to repeat what was said, you couldn't remember. Do you think failure to listen to others gives them the impression that you're not interested in them?

Because confrontation is specific, direct, and not evaluative, it can affect clients directly. Some individuals cannot easily accept another's description of reality, especially if that reality is not complimentary. And, in some cases, clients may tend to react in a negative manner regardless of the feedback. That, of course, is the client's choice. However, if confrontation occurs in a caring way, clients are likely to perceive the helper's positive intent. Unlike a casual social friend, the effective helper is seen as a concerned person who does not avoid confronting the client's reality, even though the facts may be difficult to accept.

Confrontive feedback must be well timed. The effective helper confronts the client when the client is ready for it. If a good working relationship exists between the helper and client, the most useful time for feedback is immediately after the behavior has occurred. Sometimes the helper and client will be discussing or processing a specified topic. As part of the client's behavior, something is mentioned that is definitely relevant to the concern, but to confront the client *at that moment* would be out of place and prevent systematically exploring the topic. In such a case, the effective helper makes a mental (or written) note of the behavior and brings it up at a more appropriate time.

An invitation to consider change

Tamminen and Smaby (1981, p. 42) view confrontation as "an invitation and a challenge to examine, modify, or control some aspect of one's behavior." To be most effective, the helper assists clients to be aware of their specific and generalizable patterns of behavior. Frequently, confrontation results in the client's awareness of behavior and circumstances that may have become so *routine* that they are not recognized by the client. Accurate client awareness of current behavior is the first step in the process of change. As such, confrontation invites the client to consider whether certain kinds of change are desirable or essential. If the client engages in interrupting, arguing, inappropriate or distracting mannerisms, and the like, videotaping or audiotaping such behavior can be extremely valuable as feedback. Or, lacking recording equipment, helper simulation of the client's behavior can provide another form of feedback.

There should be no doubt that the reason for confrontation is to assist the client rather than merely to express the helper's personal feelings. The helper should not vent personal feelings just to get them out of the way. The helper confronts the client in a *nonjudgmental* manner. Confrontation is aimed at helping the client to look at reality and consider alternative ways of dealing with personal discrepancies in behavior and the situation.

Confronting discrepancies in behavior

Sometimes the client will have incorrect information about events. This incorrect information can trigger thoughts, feelings, and actions that are counterproductive or destructive. In some cases, merely giving the client accurate information can be effective in changing associated behavior. In other situations, however, there are discrepancies, or incongruencies, in one or more dimensions of the client's behavior. In such cases, confrontation usually serves to make the client aware of discrepancies in combinations of thoughts, feelings, and actions, as illustrated next. In the following examples, notice the descriptions of specific behavior are made without any evaluation and judgment on the part of the helper.

1. *Discrepancies between thoughts and feelings*:

- HELPER: You say that talking about your boss doesn't upset you, yet your voice was shaky and your fingers were digging into the arms of the chair while you described her behavior.
- HELPER: You said that your spouse's cancer won't change things, yet your cracking voice [or tears, clutching the sheets, grasping the arms of the chair, or whatever] suggests otherwise. Could you tell me about your feelings?
- HELPER: You say you'll be able to get along all right with the help of food stamps, but your tears suggest that you're still frightened that you've found yourself in the position of needing to depend on others to get by.

As these examples show, discrepancies often exist between what a person thinks or says and how that person feels. Identification of specific behavior that indicates differences between thoughts and feelings can help identify concerns that may be hidden from the client. Awareness of nonverbal aspects of the client's behavior is useful in providing clues to these hidden concerns.

2. *Discrepancies between thoughts and actions*:

- HELPER: You said you would not talk to your son any more about cleaning up his room, yet in the last week your chart shows that you got into arguments about his room on three different occasions. How does that relate to your goal of wanting to communicate more positively with your son? [See Chapter 17 on charting behavior.]

- HELPER: You've told me you really want to do well in your work so that you can be promoted, but being late for work three times this week doesn't fit with what you say you want.
- HELPER: You say you're overworked and want to get help in your business, yet during the last two weeks, you haven't taken any steps to do so. There seems to be a difference between what you say and do.

As the adage states, Talk is cheap! Clients often have the best of intentions. They say one thing but never seem to get around to doing anything about it. Confronting the client with the difference between thoughts and actions can serve as a basis for serious interaction about what that person really wants to do and specific ways to take action.

3. *Discrepancies between feelings and actions*:

- HELPER: You said that separating from your spouse no longer bothers you, yet your tears suggest otherwise. Help me understand your feelings.
- HELPER: You claim you're no longer angry about being passed over for the promotion; however, your lack of sales and increased drinking indicate that something is bothering you.

Often, cues related to emotions or feelings can alert the helper to discrepancies between the client's feelings and actions. By alerting the client to nonverbal indicators of feelings (speech patterns, eye contact and facial expression, gestures, body position, etc.), the helper can accentuate important differences between actions and feelings with which the client may be struggling.

4. *Discrepancies in combinations of thoughts, feelings, and actions*:

- HELPER: When you hit her [action], it seems that you really didn't care [feeling], but now you say you really love her [thought]. Help me sort out these different behaviors.
- HELPER: You said you really wanted to finish writing the book [thought], but you haven't written anything for over a week [action or avoidance]. How do you *feel* about this situation?

Thoughts, feelings, and actions that do not match each other are incongruent. Discovering incongruencies in the components of a client's behavior alerts the helper to unresolved or potential problems. Confronting the client with conflicting aspects of behavior—in a caring way—can lead to increased awareness and acceptance of reality. Based on this new awareness, the client then can choose what to do about these specific aspects of behavior.

Confrontation can be a very useful, constructive technique throughout the helping process. Leaman (1978) cites Berenson and Mitchell (1974), who identify five major types of confrontation, categorized according to their purpose. These are to

1. Focus on discrepancies in the client's behavior.
2. Clarify misinformation or misconceptions.
3. Identify pathological patterns.
4. Motivate the client to constructive action.
5. Emphasize the client's resources and strengths.

The client quadrant

It is important to emphasize that what "reality" is for the client may be different than it is for others, including the helper. The *client quadrant* is a way of looking at how the client may perceive reality and how it can be similar to or different from reality as seen by others (Figure 9.1). The client quadrant is a two-dimensional matrix that consists of how behavior is seen by others and the client. The client and others may see any specified behavior as "no problem" or as a "problem." The interaction between how the client and others view behavior results in four major ways of viewing any behavior as described next in Quadrants I to IV.

Quadrant I: Neither the client nor others see a major problem. Typically, the client may seek information, testing, assessment, or appraisal of interests, skills, and abilities to engage in the most effective route to personal growth and development. Such a client is usually self-referred or enters into a helping relationship voluntarily to enhance personal and/or career development.

Quadrant II: In this area, the client experiences a personal concern that is known to self but often not known to others and certainly not regarded as a problem for others. This individual may be struggling with inner concerns, competing values, questions that relate to life-style, interpersonal difficulties, and other areas where there are discrepancies among thoughts, feelings, and actions. Such an individual may be painfully aware of personal concerns about which others do not know or which others do not perceive as a problem. Usually this client is self-referred to a helper.

Quadrant III: In this area, the client does *not* see a problem, but other people do. The client is content to live in the present manner, but others believe that aspects of the client's behavior (thoughts, feelings, and actions) are unacceptable. The client may have different values, life-style, or habits that others regard as a problem. The client's behavior may often conflict with local norms. This can be seen, for example, in a book that is banned in one culture but widely distributed in another, a life-style largely frowned on in one community but generally accepted in another setting, and so on.

With behavior involving substance abuse, domestic violence, sexual offenses, and the like, the individual's behavior may be extremely discrepant from others' ideas of right and wrong, besides possibly being unlawful, unethical, or immoral. In many such situations, confrontation is difficult because a major discrepancy exists between how the client sees the action and how others see it. As a baseline for any kind of helping relationship, the client must agree that the consequences of behavior necessitate change. Becoming aware of and confronting these discrepancies in perceptions of thoughts, feelings, and actions between self and others is a place to start. The

Behavior as Seen by the Client	Behavior as Seen by Others: No Problem	Behavior as Seen by Others: Problem
No Problem	I *Developmental Client* Seeks information, assessment, or confirmation of interests, skills, aptitude, and abilities; explores areas of personal and career growth and development (self-referral or referral by others).	III *Reluctant Client* Client does *not* see a problem, but others do. May be involved in abusive behavior to self and/or others. Has different values, life-style, or perception of reality than others (referral by others).
Problem	II *Personal Concern Client* Client sees concerns known only to self. Others are not aware of a problem with the person's behavior (self-referral).	IV *Consensus Client* Both client and others agree that there is a problem. Perceptions of reality are similar. Something must be done to change behavior of self and/or others (referral by self and/or others).

FIGURE 9.1 The client quadrant

client quadrant provides one way of conceptualizing reality. Using the client quadrant does not necessarily suggest judgments or accusatory behavior on the part of the helper. However, it may indicate value differences that must be examined by both the helper and the client. The client quadrant is at least a framework for confronting different ways of looking at behavior in specified situations.

Quadrant IV: Here the client and others perceive behavior as a problem. Both the client and others start from the *same perception*—that something must be done to resolve a problem. The helper assists the client in confronting discrepancies among thoughts, feelings, and actions as a baseline from which to view behavior and its consequences.

Shifts in the client quadrant

At various stages in the helping process, *the client's behavior and perception of reality may shift from one quadrant to another*, as illustrated in Figure 9.2.

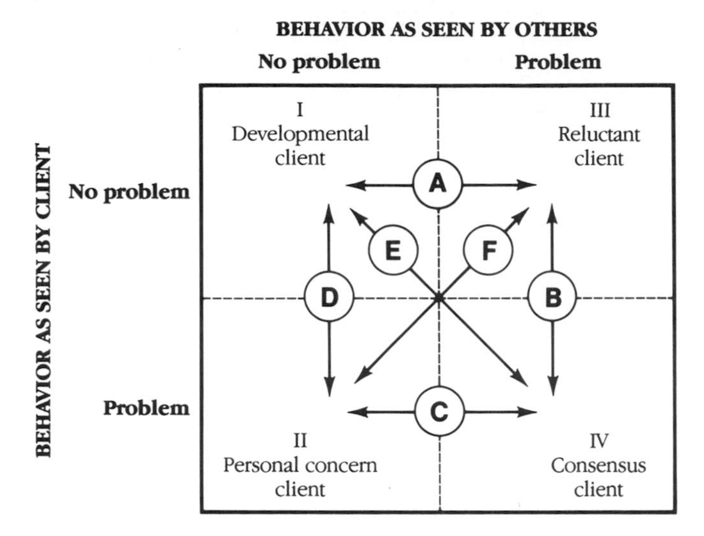

FIGURE 9.2 *A* to *F* show ways that behavior and/or the situation can shift in the client quadrant.

The following examples illustrate some of the shifts that may occur among any of the four quadrants. Letters *A* to *F* refer to some of the shifts between different quadrants.

A. A person with a preference for a gay life-style moves from a job in a small rural community, where attitudes and values of many people are intolerant or violently aggressive, to a large urban area. The client's behavior does not change, but attitudes of others with whom the person associates may be very different. The move results in a shift from Quadrant III to I because associates in the new community do not see a problem.

B. An alcoholic at first refuses to admit there is a problem (Quadrant III) but later agrees that alcohol is causing major personal, family, and career difficulties (Quadrant IV). The client then becomes actively involved in Alcoholics Anonymous, seeks medical help, and contacts a professional helper to work through related concerns.

C. A client experiences extreme personal job stress (totally unknown to others who see the individual as performing exceptionally well) and tries to cope with it (Quadrant II). As a result of continuing stress and ineffective coping, the person is arrested for shoplifting an item that could have been easily afforded. At this point, others become involved (Quadrant IV). There is a shift from a personal concern (Quadrant II) to one in which others also see a problem, such as one's family, employer, merchants, the police, the court system, and a professional helper (Quadrant IV).

D. At an early stage in dealing with personal stress, when one is dissatisfied with a living relationship or career concern (Quadrant II), the client may take positive steps to resolve these concerns before they become a major problem. For example, both partners may participate in a couples' communication workshop series, or the individual may seek career counseling long before "burnout" occurs (shift from Quadrant II to Quadrant I).

E. There might be a progression from Quadrant I (where the client seeks positive, developmental assistance and support for personal growth and development) to Quadrant IV (where areas of concern can be more fully explored with others). An example might be persons who had explored career options or couples' communication (Quadrant I) and then sought additional assistance in order to examine the impact of various kinds of changes on the individual, family, life-style, and so on.

F. A client becomes concerned that a personal life-style choice (Quadrant II) will be seen by significant others—family or employers—as a problem (Quadrant III). As an example, the client has an extraordinary career opportunity that will mean a significant leap up the corporate ladder and a major increase in salary but will also mean assuming additional administrative and management responsibilities, moving to a different location, and a host of other concerns. Assume the client does not want to take the option, but others in the family strongly encourage it. Such a situation might result in a shift from Quadrant II to Quadrant III and ultimately to Quadrant IV (where all persons concerned need to work through the situation).

As seen in the preceding examples, there may be major shifts in the client's position in the quadrant, depending on how the individual and others view both behavior (thoughts, feelings, and actions) and the situation (who's involved, where, and when). By confronting *reality* in a caring manner, the helper encourages the client to consider consequences of behavior in the specified situation.

Summary

1. Confrontation is a direct technique of helping the client increase awareness of the realities of personal behavior.
2. To actively consider behavior changes, the client must first obtain accurate awareness of reality.
3. Confrontation can be helpful if done in a caring way.
4. Effective confrontation is like useful feedback: It is specific, descriptive, well timed, and considerate of the client's needs.
5. Confrontation clarifies incongruencies and discrepancies in combinations of the client's thoughts, feelings, and actions.
6. The client quadrant can be used to confront the client's perception of reality and to explore options.
7. How the client views behavior may change throughout the helping process.
8. If a trusting relationship has been established, the client understands that the helper cares enough to confront behavior directly rather than ignoring unpleasant or difficult aspects.

Practice

1. Find a person who agrees to help you practice various kinds of confrontations. Assume that you and your coached client have been talking for about twenty minutes and that you have previously conceptualized the client's concerns.
2. Select one situation from the list of role plays, which follows these instructions. Before starting your mini-interviews, clarify specifics of the situation: age and gender of the client, setting of the interview, and any other details essential to getting started.
3. Conduct a five- to ten-minute mini-interview for each role-playing situation on the list. Record your interaction or have another person observe the interview, especially noting aspects related to your confrontation.
4. Following each mini-interview, ask your coached client for feedback about the interaction. Listen to the tape or talk to your observer and answer the questions in the "Focus Checklist," which follows the role-playing situations.
5. Then try some ten- to fifteen-minute mini-interviews in which you practice appropriately integrating the various techniques learned up to now. Use the "Summary Checklist" and record all appropriate data from these more integrated interviews. (The "Summary Checklist" and its instructions follow the "Focus Checklist.")

Role play

1. This client's *goal* is to make new friends but feels that not much progress has been made on the goal. You have determined that the client has done very little homework related to making friends during the last week. In the mini-interview, get more details of the client's behavior in this situation and confront the client about discrepancies.

2. Your client, who has always stressed the importance of obtaining a college degree, begins this interview by stating, "Since I have a job now, I guess I really don't care that much whether I go back to school." Confront the client about this discrepancy.

3. This client has been referred to you by a supervisor at work because of unsatisfactory job performance. Specifically, the complaint is that the client often smells of alcohol and has been seen going to the parking lot during work breaks. There is also an increasing record of "illness" and coming in late for work. Confront your client about this behavior.

4. The client is having problems developing and maintaining good interpersonal relationships. Assume that you have been talking with this client for the last five minutes, during which time the client has interrupted you three times. As you begin to summarize your client's interest in better relationships with others, you are interrupted again. The next time there is an interruption, confront the client about this behavior. If you are recording the

interview, you may want to use the tape as a part of the feedback and confrontation.

5. Last week your client—whose expressed goal is to learn to budget money rather than falling deeper in debt—made a commitment to begin writing down everything spent each day. At this week's appointment, it turns out that your client kept track of expenditures for only the first two days of the week and used a credit card to buy items that were not really necessary (or on budget): two stereo tapes, $16; a magazine, $4; and some new clothes, $56. Confront the client about this behavior.

Focus checklist

1. Did you demonstrate caring and concern for the client as an individual?
2. Were the specific aspects of the client's behavior described in objective terms?
3. Did you address discrepancies in the client's thoughts, feelings, and actions?
4. Did you shy away from mentioning difficult or unpleasant aspects of the client's behavior?
5. Did you avoid using emotionally loaded, evaluative terms in the confrontation?
6. What effect did the confrontation have on the client? On you?
7. What effect did confrontation have on the interview process?
8. Where is the client's position in the client quadrant?
9. What shifts must be made in the client's behavior in order to engage in a productive helping relationship?
10. Did the confrontation focus on discrepancies in the client's behavior? Clarify misinformation or misconceptions? Identify pathological patterns? Motivate the client to constructive action? Emphasize the client's resources and strengths?
11. How can you improve your methods of confrontation?

Summary checklist

As you conduct a more integrated and appropriate mini-interview, see how many of the following elements you used, and in what ways you used them.

1. Structuring the helping relationship
2. Focus on thoughts, feelings, actions or TFA combinations
3. Listening for understanding and nonverbal aspects
4. Asking open questions
5. Asking closed questions
6. Systematic inquiry
7. Clarification
8. Reflection

9. Use of silence
10. Confrontation

Instructions

To assess your interview skills, make a worksheet similar to the Behavior Summary Checklist. In the *left column* on the checklist, make a transcript of key words and phrases or complete sentences used by you and your client during the mini-interview. Then, *for each* transcript response, put a check in columns that are relevant. You might check from one to several items for each response. Ignore columns with content area that has not been discussed.

When you have completed the Behavior Summary Checklist, review it carefully. Look for patterns where you emphasized or ignored content. Then write a brief summary of the mini-interview outlining your strengths, limitations, and suggestions for improving skills.

References and recommended reading

Amatea, E. S. (1988). Engaging the reluctant client: Some new strategies for the school counselor. *The School Counselor*, *36*, 34–40.

Berenson, B., & Mitchell K. (1974). *Confrontation for better or worse!* Amherst, MA: Human Resource Development Press.

Leaman, D. R. (1978). Confrontation in counseling. *Personnel and Guidance Journal*, *56*, 630–633.

Siegel, L. I. (1987). Confrontation and support in group therapy in the residential treatment of severely disturbed adolescents. *Adolescence*, *22*, 681–690.

Tamminen, A. W., & Smaby, M. H. (1981). Helping counselors learn to confront. *Personnel and Guidance Journal*, September, 41–45.

Summary Behavior Checklist

Key words and phrases of the Client (CL) and the Helper (H)

Focus or Emphasis		
	T	
	F	
	A	
Structure		
Nonverbal aspect		
Open question		
Closed question		
Systematic inquiry		
Clarification		
Reflection		
Confrontation		
Use of silence		
Client's concern		
Goals		
Procedures		
Evaluation		
Termination		
Followup		
Strategies		

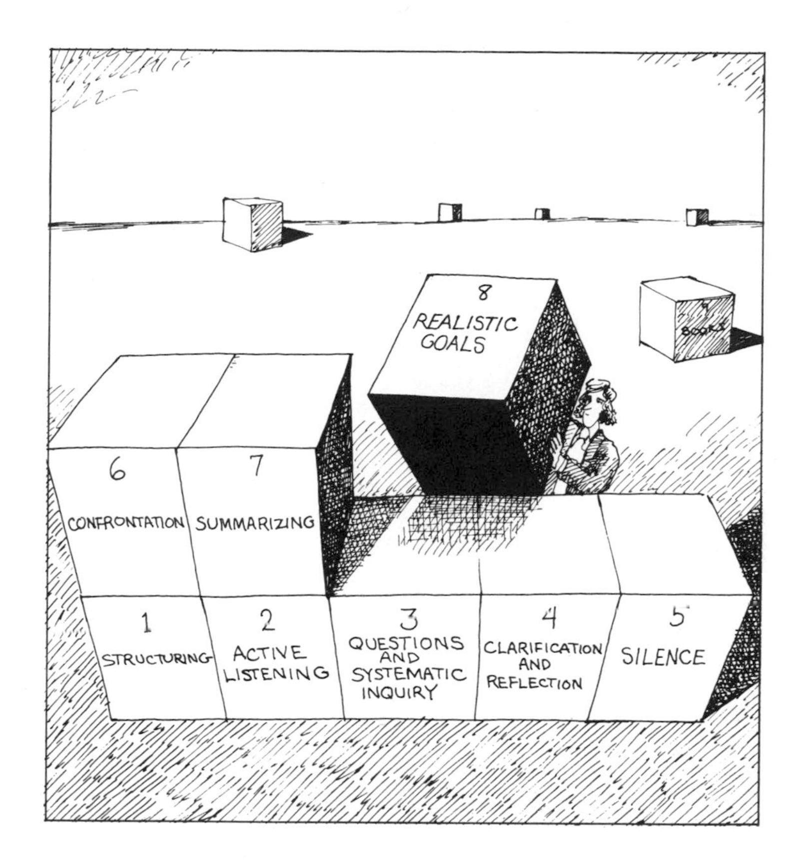

Integrating Techniques in the Problem-Solving Process

PART THREE helps readers build on the skills in Parts 1 and 2 and integrate them into an effective problem-solving process. Client concerns are identified, goals are established, and procedures are designed to help the client shift from the problem to the solution.

THREE

Identifying Client Concerns or Problems

10

Key points

1. Identifying major aspects of the concern or problem is facilitated by examining the client's thoughts, feelings, and actions in the specific situation.
2. The problem must be clearly identified at the start of the problem-solving sequence in the helping relationship. (This may be different from the initial problem the client presents when first meeting the helper.)
3. The context in which any problem behavior occurs must be understood. The cause of the problem may be outside the client (as in cases of abuse or violence).
4. Changing the environment (other people, the setting, the physical conditions) may eliminate the problem.
5. If the context or situation cannot be changed in positive ways, at least part of the problem for the helper includes helping the client make the best possible adjustment to a bad or difficult situation.
6. Client behavior can be assessed by using the TFA triangle in the interview.
7. The client's behavior pattern (TFA triad) becomes a behavior blueprint in which positive and negative thoughts, feelings, and actions are described in specific terms.
8. The TFA triad resulting from assessment is useful in describing the client's (a) behavior pattern, (b) payoffs, (c) tradeoffs, and (d) problem or concern.
9. The problem statement is a concise indicator of the client's concern or problem behavior in the specific situation.
10. With multiple problems, the priority and immediacy of each concern is indicated clearly.

IDENTIFYING client concerns or problems stems directly from assessing the client's thinking, feeling, and acting behavior in the specific problem situation (Chapter 2). Understanding important dimensions of the concern occurs as a direct result of using helping skills and techniques described in preceding chapters. Throughout the helping interview, to clarify the client's concern in the problem situation, the helper uses appropriately integrated techniques from each chapter. Furthermore, various strategies included in Part Four of this book are also extremely useful in identifying the nature of the client's concern (also see Figure 1.1).

The problem and the situation

Interaction always occurs between the person and situation or context. However, in identifying the problem, examining *antecedents* of the client's behavior is imperative (Chapter 2). In fact, the client's "problem" may not be caused by the client's behavior but may occur as a consequence of others or the situation. Situations of psychological and physical abuse or violence illustrate this most clearly. In such situations, another person or persons are major perpetrators of behavior in which the client is a victim (Clow, Hutchins & Vogler, 1990). For example, childhood pregnancy, incest, sexual abuse, date rape, and other violent behavior may occur because of the domination of another person. The result of such behavior must be worked through with the client/victim (Tieman, 1991). However, changing the physical environment, removing the client from the situation, and/or arresting others for unlawful, immoral, or unethical behavior may first be essential. If the situation cannot be changed, more appropriate avoidance or coping behavior must be taught.

At this point in the interview, the *helper's* goal is to assist the client in determining the client's concern in the problem situation. The following is an overview of important concepts related to identifying the concern or problem:

1. *Identify the problem situation* by specifying concrete aspects in which the problem occurs.
2. *Assess the client's behavior* in the situation by using the TFA triangle and connecting the points on each side of the triangle to form the TFA triad.
3. *Interpret the meaning of the behavior pattern* in the problem situation.
   a. General description of the client's behavior
   b. Payoffs/strengths for the client
   c. Tradeoffs/limitations for the client
   d. Implications of this behavior for the client

All steps are briefly outlined next in an illustration of how they are used to identify the client's concern.

Identify the problem situation

If there is no problem, there is no reason for the helping relationship to occur. The helper's objective here is *to gain some understanding of the client's concern in the problem situation.* In exploring the problem, the helper determines specific, relevant aspects such as

- *Who* is involved in the situation? Are there others (individuals or groups) who are consistently present when the problem occurs?
- *Where* do the events occur? At home, work, recreational, or leisure settings?
- *When* do the incidents happen? Early in the day, at mealtime, late at night?
- *What* is the immediacy, frequency, and severity of the problem? Might the situation occur again right away, once a week or month, every few months? Is it a crisis or relatively harmless situation?

- *What* are the unique circumstances associated with the problem behavior?

Assess the client's behavior

The client's behavior is the interaction of thoughts, feelings, and actions. We recommend that the helper use the TFA triangle with clients in the interview. The helper asks three critical questions and records the client's responses on the TFA triangle (Figure 10.1).

In reference to the problem situation,

1. Is your behavior (3) more thinking, (1) more feeling, or (2) about in the middle?
2. Is your behavior (3) more feeling, (1) more acting, or (2) about in the middle?
3. Is your behavior (3) more acting, (1) more thinking, or (2) about in the middle?

The helper then connects the points on each side. These form a closed inner triangle called a *TFA triad*. Operationally, this TFA triad is a graphic representation of the client's interaction of thoughts, feelings, and actions in the problem situation. As Figure 10.1 illustrates, the client's TFA triad is predominantly thinking and feeling without action.

Interpret the meaning of the behavior pattern

As the client looks at the pattern, the helper asks what the TFA triad means about his or her behavior in this situation. The chances are excellent that the client will describe some things that relate very strongly to the nature of the

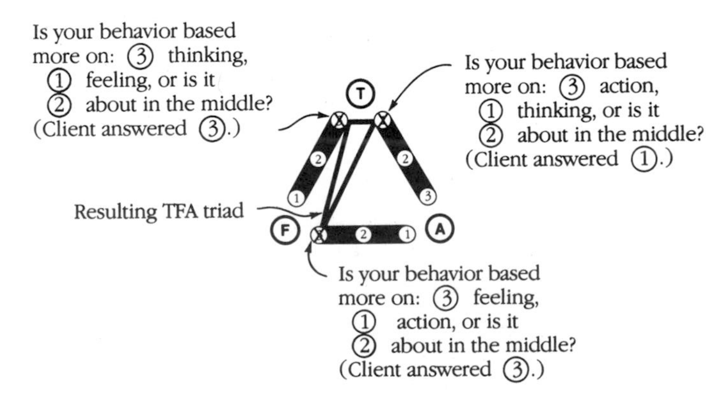

FIGURE 10.1 Three critical questions about the client's behavior in a specific situation. Responses are plotted on each side of the triangle and then connected to form the TFA triad, which shows the interaction of thoughts, feelings, and actions in the specific problem situation.

Thoughts
I can't possibly talk with others.
People wouldn't understand me.
I wouldn't know what to say.
Others would be looking for mistakes.
People would make fun of me.

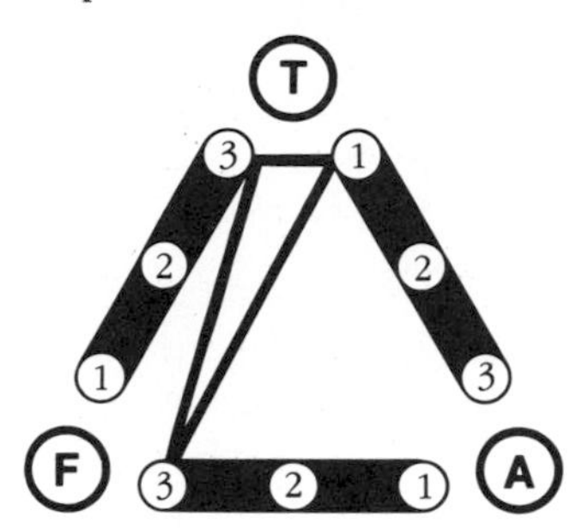

Feelings
I'm scared and worried.
I'm uncomfortable and anxious.
I get nervous and sweaty.
I get embarrassed.

Actions
I just stand off by myself.
I don't usually go where other people gather in groups.
I try to find things to do by myself.
I just listen to others.

Problem Statement or client concern:
"I'm afraid to interact with others in an unfamiliar social setting."

FIGURE 10.2 Interpreting the TFA triad results in a behavior blueprint of the client's thoughts, feelings, and actions in the problem situation.

problem behavior. To get a more complete description of the client's specific behavior, the helper continues the interpretation process by asking the client to describe very specific aspects of behavior at each point in the triangle. The questioning begins where the TFA triad is most dominant and then moves to where it is absent (or latent). In the case where two or three points are equally dominant, begin the inquiry at any point.

To illustrate, assume that the client has indicated a concern about feeling uncomfortable around others in a social situation and has completed the TFA triangle, which results in the pattern shown in Figure 10.1. The helper obtains more details starting with the *left* (T–F) side of the triangle by probing and inquiry such as that outlined below for each side of the triangle. The process is repeated with the other two sides (F–A and A–T), using a sequence from the most dominant part of the TFA triad to the absent or latent area. Complete processing the client's behavior on one side of the triangle before moving to the other side. Ask the client for specific reasons for indicating the particular point on each side of the triangle. Sample questions and examples of a client's responses are illustrated below and summarized in Figure 10.2. This results in the helper getting a *behavior blueprint* of the client's specific thoughts, feelings, and actions in the problem situation.

T–F (left side of triangle): helper inquiries

What made you say you were more thinking [or feeling, or about in the middle]?
List as many ways as you can that describe this behavior.
What were you thinking [why were you feeling that way] just prior to getting involved in the situation? Be specific.
What else were you thinking or feeling?
What emotions did you have?
Why were you feeling as you were?

Figure 10.2 illustrates what a client might say. Even though questions are asked about thoughts or feelings, the client often confuses these, failing to distinguish between thoughts and feelings. When this happens, the counselor should (1) briefly write on the triangle whether the behavior was a thought, feeling, or action; (2) let the client know what it was (thought, feeling, or action); and (3) probe for more information. Learning to discriminate between thoughts and feelings is an important part of this activity for both the client and helper.

- CLIENT RESPONSES (THOUGHTS): I can't possibly talk with others. People wouldn't understand me. I wouldn't know what to say. Others would be looking for mistakes. People would make fun of me. Sometimes I wonder what others are thinking.
- CLIENT: I was really *nervous*.
- HELPER: Okay, nervousness is a *feeling*. What were you thinking that made you so angry?
- CLIENT: I guess I was thinking that others would make fun of me and I wouldn't know what to do.

In the process of probing for specific aspects of the client's behavior in the problem situation, the helper uses *all* appropriate skills addressed earlier, such as reflection, clarification, attention to nonverbal aspects of behavior, systematic inquiry, and so forth. As a result, what emerges is an extremely rich, descriptive understanding of important dimensions of the client's behavior in the problem situation.

F–A (bottom of triangle): helper inquiries

Why do you say you were more feeling [or how were you more action-oriented, or why do you say you were about in the middle]?
How did you feel in that situation?
Can you describe these emotions [or actions] in more detail?
What did you do? What actions did you take?
How did you respond to . . . ?

- CLIENT RESPONSES (FEELINGS): I'm scared and worried. I'm uncomfortable and anxious. I get all nervous and sweaty. I get embarrassed.

Again, as the client describes details of personal behavior, the helper follows up on these responses with whatever techniques are appropriate, helping the client expand on important aspects, reflecting content and emotions, clarifying, summarizing, and so on.

A–T (right side of triangle): helper inquiries

How were you more acting [or what makes you say you were more thinking or about in the middle]?
What kinds of things did you do?
How did you respond? Be specific.
How did you get there?
What happened when you arrived at . . . ?
What were you thinking?
What ideas were going through your head?

- CLIENT RESPONSES (ACTIONS): I just stand off by myself. I don't go where other people get together. I try to find things to do by myself. I just listen to others, but sometimes I wonder what they are thinking.

Have the client summarize the major concern

When the helper has completed the inquiry about the client's behavior in the problem situation, the client should try to summarize the essence of the problem.

- HELPER: You have described specific aspects of your behavior [as represented by the TFA triad]. Look at the TFA triangle and summarize the major concern or problem as you see it.

As an alternative, the helper can summarize the essence of the concern, but it is best to have the client do it first. For the case described in Figure 10.2, the client's summary might be something like this:

- CLIENT: I guess I'm afraid to talk with others when I'm around people I don't know. I just feel anxious, and I don't know what to do.

Using the TFA method of inquiry, we find that helpers can assist clients by pinpointing specific, concrete aspects of the client's behavior while gathering more relevant data than is typical for beginning helpers.

Observations and suggestions for the process

1. *Help the client distinguish thoughts, feelings, and actions from each other.* In some cases, the client cannot separate thoughts, feelings, or actions from each other. Thoughts may be fused with feelings or actions. Similarly, feelings are fused with acts. A person might say, "What I *did* was to get *angry*." However, anger is a feeling. As a result of anger, one may do certain things

(act). It is essential to recognize that feelings are different from actions. For example, often in the case of violent behavior, the abuser equates feelings (of anger, jealousy, or rage) with actions, failing to recognize that one can feel a certain way but that actions taken can be independent of one's feelings. Many people experience anger, but it does not have to lead to violent or abusive behavior. By itself, this awareness can be extremely helpful.

2. *Take notes.* As the client is talking, jot down specific cues related to the client's behavior. These notes are invaluable. They provide a behavior blueprint of specific aspects of the client's behavior in the problem situation (see Figure 10.2). This behavior blueprint provides exactly the kind of data you need to assist the client in knowing specific behavior that must be changed if the problem is to be resolved. Look at Figure 10.2 and note the irrational *thoughts* that need to be changed, *feelings* that are counterproductive, and *actions* that indicate what the client needs to do to shift into more appropriate and effective behavior.

3. *Revise the triad.* Sometimes it may be necessary to revise the triad after this process to more accurately reflect the individual's behavior. For example, the client may initially indicate mostly thinking when it is really feeling, or mostly feeling when the behavior is mostly action, and so on. This activity provides valuable clues to help the client see behavior more accurately and in ways that the helper can use to assist the client.

4. *Repeat the process as often as necessary.* Throughout the helping relationship, the TFA triangle assessment can be helpful in understanding the client's behavior at different times and in different circumstances. For example, in working with a spouse abuser, it may be desirable to determine the behavior occurring early in the day, several hours before, or in the moments just preceding the abusive act (Clow, Hutchins & Vogler, 1990). Knowing the client's behavior, as reflected in TFA triads, that occurs at various times in a sequence of events may be very helpful in resolving problems and preventing future problems.

Multiple problems can be worked through in the same way as a single problem, using the TFA procedure. The helper then assists the client in ranking the various problems in terms of immediacy and priority. In most cases, actual work on the problem starts with whatever causes the greatest immediate concern for the client. The helper is especially sensitive to aspects of the situation that can be alleviated right away with accurate information or referral to other resources, such as medical personnel, shelters, or child protective services. Overriding or more deeply entrenched problems can be dealt with over a longer period of time. Generally, the sequence of working with concerns is last (or most recent) things first.

In assessing the client's behavior using the TFA triangle, the helper works with the client in determining four major aspects that are relevant to the helping process. Below each one is an example of that concept for the client who has difficulty talking and interacting with others in a social situation (Figure 10.2).

1. What is the *general description* of the client's behavior in the problem situation?
   - CLIENT DESCRIPTION: The TFA triad indicates one who is likely to think, analyze things in detail, and feel strongly about them. Thoughts and feelings are likely to be interactive and influence each other. If feelings are positive, thoughts will also be positive, whereas if either thoughts or feelings are negative, they are likely to influence each other.
2. What are the *payoffs*—strengths or what the client is gaining from the particular behavior in the situation, as reflected by the triad—of the current behavior?
   - CLIENT PAYOFFS: The client has ideas and plans. There may be initial enthusiasm for projects. The client probably conceives ideas and may have strong feelings about them. The client may relate relatively well to others who are already known in a one-on-one situation. The client is sensitive and aware of personal discomfort and the discomfort of others in social situations.
3. What are the *tradeoffs*—limitations, negative aspects, or disadvantages of the client's behavior—in the specified situation?
   - CLIENT TRADEOFFS: The client rarely interacts with others unless the individuals are already known. The client stands off and avoids participating in events, which limits opportunities. The client's social skills cannot be developed because of lack of practice in actual social situations.
4. The *problem statement* is a concise description of the problem or concern and can be made by either the client or helper.
   - PROBLEM OR CONCERN: The client *feels* anxious or uncomfortable with people he or she doesn't know, *thinks* about embarrassing events that might occur, and *stays away* from situations where social interaction might occur.

A clear statement of the problem reflects relevant aspects of the client's thoughts, feelings, and actions in a specific situation. This is critical because a good problem statement leads directly to specifying goals. As seen in item 4, the concerns of the client clearly imply changes that can be made in thoughts, feelings, and actions.

Summary

1. The client's behavior in a problem situation must be clearly identified before advancing to other aspects of the problem-solving process.
2. The client's behavior can be assessed with the use of the TFA triangle, which leads to a TFA triad.

3. The TFA triad is a graphic representation of the interaction of the client's thoughts, feelings, and actions in the problem situation.
4. The TFA triad results from asking three critical questions about the client's behavior in the specific problem situation, which are plotted on each side of the TFA triangle:

   Left side: Is your behavior more thinking, more feeling, or about in the middle?

   Bottom: Is your behavior more feeling, more action-oriented, or about in the middle?

   Right side: Is your behavior more action-oriented, more thinking, or about in the middle?

5. The helper asks for specific descriptive information about the client's behavior, beginning first where the triad is the most dominant and moving to where it is least dominant.
6. The behavior blueprint documents positive and negative aspects of the client's thoughts, feelings, and actions in the specific situation.
7. Following the behavior blueprint, the helper and client review the behavior pattern, payoffs, and tradeoffs, and develop a specific problem statement.
8. The problem statement should be a concise description of the client's behavior (thoughts, feelings, and actions) in the specific situation.
9. With several different problems, the client and helper determine the relative priority and immediacy of each.
10. Generally, the sequence of working on concerns is *last* (most recent) *things first.*

Practice

1. Find a person who agrees to be your coached client and help you practice identifying concerns by participating in a number of ten-minute mini-interviews.
2. Select a role-playing situation, which follows these instructions. Before starting to role-play, clarify specifics of the role-playing situation: the client's age and gender, the setting, and other details necessary for the interview.
3. Assume you have already been in the interview for about thirty minutes. Record your interaction or have an observer take notes. (a) When you start the interview, use the TFA triangle and ask the three critical questions. (b) Complete the TFA triad. (c) Determine the meaning of the client's behavior as illustrated by the triad so that you have a descriptive behavior blueprint similar to that in Figure 10.2. (d) *In the last minute,* you or the client should summarize major aspects of the client's concern in a statement of the problem. Give this summary in one or two concise statements, not exceeding one

minute in length. (*Note*: To do all this in ten minutes will seem a bit rushed, but with practice it will come easily.)

4. Following each mini-interview, listen to the tape or ask the observer for feedback. Then answer questions in the "Focus Checklist."
5. When you have completed these tasks, try other ten-minute interviews in which you practice appropriately integrating various techniques learned up to now. Use the "Summary Checklist" at the end of this chapter and record all data from these more integrated interviews.

Role play

1. Your client has been increasingly concerned about sexual identity and relationships with others, at times feeling a very strong attraction to certain members of the same sex. During the interview, the client expresses confusion about these desires and their meaning and explains that most of her friends have no idea that these feelings exist.
2. Thinking about career interests, your client is strongly motivated toward an area that is traditional for members of the opposite sex. The client has concerns about the worth of doing preparation for entry into that field and whether it would be too much of a hassle.
3. Faced with a strong possibility of undergoing surgery, the client seeks your assistance, saying:

   "I just can't think straight anymore. The whole thing has really caught me off guard. What if the tumor is malignant? I mean, I just never thought about such possibilities before. Now I can't think about anything else."
4. Your client reports to you that a job held for eleven years is being eliminated in a major company reorganization. A single parent with two children in high school, your client is aware that there is no similar kind of job possibility in the area and is afraid of what lies ahead.

Focus checklist

When you have completed a role-playing situation and are listening to your taped interview or getting feedback from an observer, assess *your* performance with the appropriate items below:

1. What is the client's behavior blueprint in terms of thoughts, feelings, and actions in the problem situation?
2. Did you describe the essence of the client's problem behavior or concern in the specific situation?
3. What are payoffs and tradeoffs (positive and negative aspects) of the client's behavior blueprint in the problem situation?

4. Did you prioritize the client's concerns in terms of immediacy or severity?
5. Are there patterns or themes that run through the interview?
6. What are the client's personal values, and how do these influence the situation?
7. Are there special concerns of the client that relate to gender, culture, or socioeconomic issues?
8. Who are other important people in the situation, and how does their behavior affect the client?
9. Are there other aspects in this situation that are important in understanding the client's concern?
10. Did you end the interview with a concise statement of the problem?

Summary checklist

How many of the following elements did you use, and in what ways did you use them during your more integrated mini-interviews?

1. Structuring the helping relationship
2. Focus on thoughts, feelings, actions, or TFA combinations
3. Listening for understanding and nonverbal aspects
4. Asking open questions
5. Asking closed questions
6. Systematic inquiry
7. Clarification
8. Reflection
9. Use of silence
10. Confrontation
11. Client concerns

Instructions

To assess your interview skills, make a worksheet similar to the Behavior Summary Checklist. In the *left column* on the checklist, make a transcript of key words and phrases or sentences used by you and your client during the mini-interview. Then, *for each* transcript response, put a check in columns that are relevant. You might check from one to several items for each response. Ignore columns with content area that has not been discussed.

When you have completed the Behavior Summary Checklist, review it carefully. Look for patterns where you emphasized or ignored content. Then write a brief summary of the mini-interview outlining your strengths, limitations, and suggestions for improving skills.

References and recommended reading

Clow, D. R., Hutchins, D. E., & Vogler, D. E. (1990). Treatment for spouse abusive males. In S. M. Stith, M. B. Williams, & K. Rosen (Eds.), *Violence hits home.* New York: Springer.

Cormier, W. H., & Cormier, L. S. (1991). *Interviewing strategies for helpers* (3rd ed.). Pacific Grove, CA: Brooks/Cole.

Summary Behavior Checklist

Key words and phrases of the Client (CL) and the Helper (H)

T F A

Focus or Emphasis	
T	
F	
A	
Structure	
Nonverbal aspect	
Open question	
Closed question	
Systematic inquiry	
Clarification	
Reflection	
Confrontation	
Use of silence	
Client's concern	
Goals	
Procedures	
Evaluation	
Termination	
Followup	
Strategies	

Establishing Realistic Goals

11

Key points

1. Goals relate to the client's thoughts, feelings, and actions in the *future.*
2. Goals must be positive and realistic for the client.
3. Achieving goals will result in changes, which will resolve the concern or problem, in the client's thoughts, feelings, and actions.
4. The TFA triad helps determine (a) presence or absence of behavior and (b) positive and negative behavior, which provides clues for goals.
5. Five guidelines for stating goals are
   a. Goals must be stated so that they can be achieved.
   b. Conditions of performance must be specified.
   c. Consequences of goal achievement must be understood.
   d. Larger goals must be broken down into parts.
   e. Each goal must be measurable in some way.
6. The client is responsible for achieving goals set in the helping process; the helper is responsible for ensuring that goals are realistic.

AFTER laying the foundation by identifying the client's concern or problem, the helper assists the client in establishing realistic goals. Specific positive and negative thinking, feeling, and acting dimensions of the client's behavior are already known as a result of previously constructing a *behavior blueprint* from the TFA triad. As seen in Chapter 10, a good problem statement clearly implies goals that can help resolve the client's concerns. The objective for this part of the helping interview is to establish realistic, achievable goals, which will resolve the concern or problem and result in more productive, positive, and satisfying behavior (Krumboltz, 1988; Sampson & Krumboltz, 1991). The major question to be explored now is, What will resolve the problem behavior? Or, what would be a better alternative than the client's current behavior? Both the client and helper explore such a question.

The TFA triad is present or absent

In examining the client's behavior, recall that the TFA triad is drawn inside the TFA triangle. The triad is a graphic representation of the dynamic interaction of client thoughts, feelings, and actions in the problem situation. Goals can be partially determined by examining the triad in relation to the TFA triangle. In Figure 11.1a, the client's triad shows a feeling–acting pattern without thinking; Figure 11.1b shows a strong action triad with a balance between thinking and feeling; and Figure 11.1c shows a relative balance between thinking and feeling but without acting. *Presence* of the triad indicates strengths of the client's behavior in the specified situation.

Operationally, strength of one's behavior refers to where one's pattern is most dominant within the TFA triangle. This pattern indicates the greatest probability of client behavior in the problem situation. However, the helper is

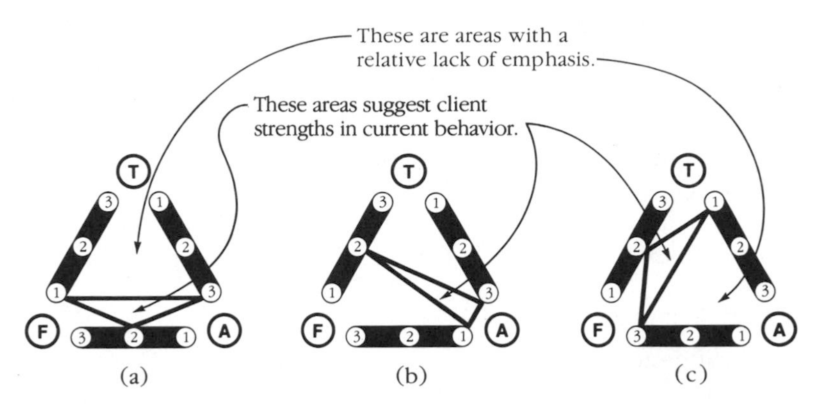

FIGURE 11.1 Three different TFA triads illustrating (a) a feeling–acting orientation, (b) an action orientation with a thinking–feeling balance, and (c) a thinking–feeling orientation with a midpoint balance between thoughts and feelings.

aware that specific behaviors (thoughts, feelings, and actions) are likely to be positive or negative.

Absence of the triad from a portion of the TFA triangle suggests that the particular area is *not* a strength of the client in the situation. Hutchins and Vogler (1988) believe that, if a triad is relatively absent, most people have a latent (undeveloped) ability to engage in behavior in the area. The triad is useful because the helper and client examine where the triad is present *and* where it is absent.

Figure 11.1 illustrates client triads in three different problem situations. The pattern in Figure 11.1a is without thinking; Figure 11.1b shows a balance between thinking and feeling with dominant actions; and action is absent in Figure 11.1c. Each triad suggests potential behavior strengths/payoffs and limitations/tradeoffs. In looking at potential goals that will resolve the client's concern, three areas relate to the client's current behavior: where the triad is present, where the triad is absent (or latent), or combinations of these as Figure 11.2 shows.

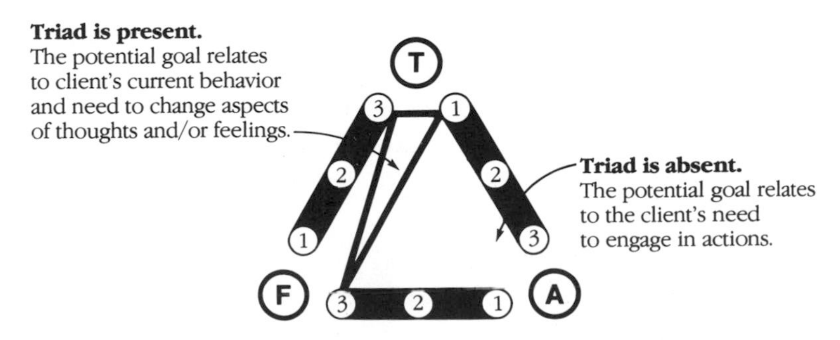

FIGURE 11.2 The client's triad and potential goals. Goals may relate to where the client's triad is present, absent, or combinations of these.

1. *Where is the triad present?* Critical questions lie in exploring aspects of the client's present behavior in the problem situation.

Thoughts

Does the client have accurate information?
Are the client's thoughts rational and soundly based?
Are the client's thoughts irrational or inappropriate for the situation?
What behavior does the client need to continue, stop, or learn to do?

Feelings

Are feelings and emotions congruent with thoughts and actions?
Are feelings appropriate for the situation and time?
What behavior does the client need to continue, stop, or learn to do?

Actions

Are the client's actions congruent with thoughts and intent?
Are actions appropriate and effective in achieving the client's goals?
What behavior does the client need to continue, stop, or learn to do?

2. *Where is the triad absent (or latent)?* The primary question here addresses the client's limitations or weaknesses in specific thinking, feeling, or acting areas. In what thinking, feeling, or acting areas does client behavior need to be strengthened or new behavior learned? In Figure 11.1a, the triad is absent in the thinking area, suggesting the possibility that the client does *not* stop to consider options or consequences of behavior before acting. In Figure 11.1c, the triad suggests the person may dwell too long on thoughts and feelings, perhaps without taking action.

Both areas where the triad is *present* and *absent* may be targets for client goals. The client in Figure 11.1c may need to stop irrational, counterproductive thoughts while learning more assertive behavior, as in the case of one who has stage fright or one who mentally rehearses negative self-talk ("I know I'll make a fool of myself") and avoids participating in events with others who are already known (see Chapter 19 on imaging and positive self-talk).

Achieving goals means changing behavior

To resolve a problem, the client must change behavior. Goals relate to what the client will be doing in the future to overcome, reduce, or eliminate concerns or problems that currently exist. This means that the client will deliberately change combinations of behavior in some of the following ways:

1. *Change thoughts*: The client changes from irrational, unproductive, self-destructive, negative thinking to more positive thinking about self and others. The client learns to accept self and aspects of situations that cannot be changed.
2. *Change feelings*: The client changes worrisome, hostile, depressive,

angry, or negative feelings to more positive feelings and emotions about self and others.

3. *Change actions*: The client changes from displaying destructive, inappropriate, spontaneous, or counterproductive actions to more positive, constructive, socially appropriate, and thoughtful goal-oriented actions.[1]

Guidelines for stating goals

The second step is to establish realistic, achievable goals. This section contains five important guidelines for stating goals. First, state goals so that they can be achieved. Second, specify conditions under which goals will be performed. Third, the client must understand the consequences of achieving goals before committing to them. Fourth, if the goal is too large, break it down into achievable parts. Finally, make goals measurable. Each of these is discussed in some detail.

1. *Stating achievable goals*. Goals must be stated in *positive terms* that define *what the client will be doing*. The client needs a positive idea or image of what will be done (Glasser, 1981; Maultsby, 1984). It is *not* effective to say "I will refrain from yelling," "I will stop smoking," "I will not watch TV," or "I will not procrastinate." To change behavior, the client must have a positive mental image of what will be done. The more concretely or specifically the client can picture or imagine the event, the better. Thus, the negative ideas are better stated as "I will *count to ten before I say something,*" "I will *go outside and work,*" "I will *read a chapter in a book,*" or "I will *start doing my homework immediately after lunch every day*."

It is helpful to use the TFA framework to formulate the goal. When the goal is achieved, what will the client be doing in terms of thoughts, feelings, and actions? For example, in Figure 11.1c, the major goal is that the client *think* positive thoughts, *feel* relaxed, and engage (*act*) in conversations with others.

2. *Narrowing from vague to specific*. Especially in the beginning stages of the helping relationship, the client may have only a vague, general idea of personal goals. Such general goals provide a starting point for exploring, probing, and defining more specific goals as illustrated in the sequence below:

- HELPER (assume that the client's feelings about the work situation have been explored in some detail): What do you think you would like to do about this work situation that you find so impossible?

[1]Thinking, feeling, and acting components of behavior are interactive and overlapping. Debate continues about what dimension of behavior occurs first, whether thoughts or feelings are primary (Lazarus, 1984; Zajonc, 1984). The position taken in this book is that the client's thoughts, feelings, and actions can be used in a *practical* way in the helping relationship. Any of the three dimensions of behavior can be regarded as primary in a specified situation and used to assist in the change process.

- CLIENT: I don't know. I just feel as if everything is closing in on me. I have so much to do at work that I don't know where to start. I'm frustrated and disorganized and everything is mixed up.
- HELPER: If you could change two things to make your work less frustrating and more manageable, what is it that you would like to do?
- CLIENT: I guess I don't like the mess and disorganization.
- HELPER: Your facial expression communicates your frustration. Tell me, specifically, if things were more positive for you, what would they be like?
- CLIENT (summarizing after about twelve seconds of silence to ponder): Well, I guess I'd be able to *prioritize activities* in terms of their importance and *organize my time* better.
- HELPER: Good! You've mentioned two specific tasks [goals] you could do to help get control of your work.

In the first part of this exchange, the client described the frustration and disorganization. This was a *poor goal* because it (1) described what the client did not like (the frustration and disorganization), (2) offered no guidelines for what the client would be doing, and (3) was negative ("I don't like . . ."). As the interview continued, the helper assisted the client in focusing on *positive* and *specific* kinds of actions that could be done to resolve the concern.

Specific, positive goals provide useful ideas about *what the client will be doing*. Identifying the client's concern or problem (Chapter 10) often provides a *description of the current situation* and specific, concrete data that suggests what might need to be done. Goals provide direction for the client.

3. *Specifying the conditions.* What is the situation or context in which the new behavior will be taking place?

- *When* will the behavior occur?
- *Where* will the behavior take place?
- *With whom* or *in what circumstances* will the behavior occur?

Adding these three elements makes the client's goal more specific. The goal now becomes, *When* I meet other people I don't know (*who*), I will *think* positive thoughts, *feel* relaxed, and engage in conversations (*act*) that are appropriate for the time available (*circumstances*).

4. *Examining consequences.* Before the client starts to work toward goals, examining the *positive and negative consequences that might result if the goals were achieved* is crucial. The client should also think about what might happen in the *process* of working toward potential goals. Nothing happens without consequences or tradeoffs of various kinds. If you decide to do one thing, other things have to be given up or modified in some ways. The following examples show what can happen when one acts to achieve a goal *without* considering positive and negative consequences that are associated with it.

- GOAL: Sam, who had a responsible position, good salary, and good associations in the community, resigned his job to go to graduate school and complete a doctorate to attain a higher administrative position in the local school system.
- CONSEQUENCES: The family was unhappy with a move from their home town to the university. All savings were used by the end of the first year, and by that time Sam and his wife had separated. He was so emotionally upset that he dropped out of school to take a temporary job and salvage what he could of the family.

- GOAL: Samantha wanted to be more assertive with people so she attended a series of assertiveness training workshops.
- CONSEQUENCES: As a result of the workshops, she became much more assertive, behaving very differently from the rather timid way she had always been. Now, however, her husband regards Samantha as a different woman from the one he married, and there is a great deal of distress and tension in the family.

- GOAL: Ron had been an excellent automobile mechanic but wanted to become a maintenance supervisor because of additional pay and more responsibility.
- CONSEQUENCES: Ron applied for the job and was promoted to supervisor but is now frustrated much of the time. In the past, he had been an excellent mechanic who usually worked alone. The new position requires hiring and firing personnel and keeping up with a mountain of paperwork, all of which resulted in great personal stress. During the last year, he has gained twenty-seven pounds, has developed an ulcer, and has become irritable and extremely impatient with co-workers.

All too frequently, people go charging blindly over the fence with the unexamined belief that the grass is greener on the other side or that opportunities lie elsewhere. What happens is that they play the mental game "If Only I Could" in which they have an idealistic, unrealistic picture of how good things will be. For example,

If only I could be in that other person's position—never realizing that the other person puts in about twice the work hours of the client and does many unpleasant tasks that the client would not like to do.

If only I could have more education—not discovering until later that there is no possibility of being hired in that locale because the client has become overqualified and the organization is hiring people with less training for lower salaries.

If only I could move somewhere else, such as California, New York, Canada, Europe, or Australia, I'd find work—without realizing that the unemployment lines may be even longer, opportunities more limited, or job requirements more challenging in other places.

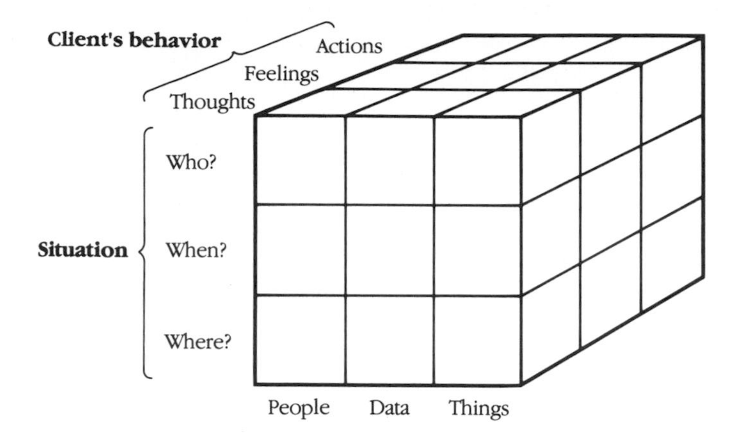

FIGURE 11.3 A chunking cube to help organize or combine content or information that relates to the client's behavior, the situation, and people, data, and things, which may be involved in goal-related behavior. (Thanks to the 1990 summer school techniques class.)

The helper assists the client in examining realities of what *might happen* should the desired goal be achieved. Information obtained through magazines, newspapers, occupational handbooks, people, and other resources can help find the facts of what is involved in a potential situation that interests the client. The helper assists the client in considering both positive and negative aspects of any situation related to goals before the client starts to work on them. One helpful technique is to construct a balance sheet on which to record positive and negative potential consequences. The format in Figure 11.3 can be used and adapted to help the client consider the positive and negative potential consequences in several broad areas of concern. Other areas important to each client should be added to this list. Use other words as organizers to help combine or "chunk" content as you help the client examine options (see the chunking examples in Chapter 6). For example, three organizers are (1) the client's behavior (thoughts, feelings, and actions); (2) who, when, and where; and (3) people, data, and things. By combining such content areas, all kinds of potential consequences emerge, which help the client explore options and consequences for specific topics. For example, along the situation side of the cube, the helper can assist the client in looking at potential content such as

- *Personal aspects*: Client thoughts, feelings, and actions about self.
- *Social and family aspects*: Formal and informal interactions with people in the family and community.
- *Education and training aspects*: Formal and informal kinds of education needed to achieve goals.
- *Job and career aspects*: Formal and informal, full- and part-time, paid and volunteer experiences.

- *Leisure aspects*: What the client does or would like to do in leisure time and how this affects other aspects of one's life (McDaniels, 1989).
- *Other aspects*: Any other areas of the client's life or anticipated experiences that must be explored before making a commitment to particular goals.

The chunking concept can be helpful in learning to combine several different kinds of content. This content can be explored with the client to determine if all critical areas of potential behavior have been explored.

4. *Breaking down larger goals into parts*. Goals must be broken down into realistic components or subgoals, if they are too broad. The helper's goal is to ensure that the client's goals are realistic and achievable. For example, if the client's general goal is to interact effectively with others in a social setting, the helper assists in breaking down this large goal into realistic, achievable parts.

General goal: To interact effectively with others
Parts or subgoals

1. To learn nonverbal aspects of interaction
2. To tell others about selected personal interests without appearing self-centered
3. To inquire about others' interests without prying
4. To interact in a conversation

By breaking down a larger goal into essential parts, the client can more easily master the parts to achieve the overall objective. To help the client achieve the subgoals, each of the preceding four areas could then be broken down into specific steps or procedures. Chapter 12 describes *how* to assist the client in achieving goals and subgoals.

5. *Making goals measurable*. Normally, if goals are stated positively, have specified conditions for their achievement, and are broken down into achievable parts, *goals will be measurable or observable in various ways*. The helper works with the client to agree on what evidence would be adequate to confirm that the goal had been achieved. Three components of behavior can make the confirmation:

1. Changes in *thinking*: As evidence of changes in thinking, the client records negative and positive thoughts about a specified topic—for example, negative or "self-downing" thoughts versus constructive thoughts associated with specific events.
2. Changes in *feeling*: As evidence of changes in feelings, the client records feelings and emotions when certain events occur—for example, changes from feelings of anger, jealousy, and anxiety to feelings of relative comfort and relaxation.
3. Changes in *acting*: As evidence of changes in acting, the client records what actions are taken at specific times—for example, one client wrote, "I went home, put on my tennis shoes and sweats, and

went for a brisk walk instead of going to the ice cream shop for a malted milk."

Helpers today must be accountable for their actions on behalf of clients. It is not acceptable to see clients, *presumably assist them,* and send them on their way. We strongly believe that the helper *must* be increasingly able to document effectiveness in helping clients live more effectively. With increasing demands for time and money, it is imperative that professional helpers demonstrate that their work makes a positive difference to their clients. At a minimum, by specifying behavior change, the helper must be responsible for assessing the results of the helping process for each client. As evidence accumulates across many individual clients, the helper establishes a record of accountability as a professional. (See Chapter 13 on termination and follow-up.)

Client responsibility for personal change

Goals in most helping relationships are usually client-initiated rather than helper-initiated. The client must assume ownership of and responsibility for personal goals. The helper assesses the client's motivation, desire, or commitment to change behavior. (We discuss more fully assessment of one's motivation for change in Chapter 12.) A client who does not really want to change behavior must be confronted with this reluctance when it occurs in the helping process (see Chapter 9 on confrontation). Among other matters, the helper may need to provide additional structure to the interview by clarifying the purpose and expectations of the relationship:

- HELPER: You will recall that the purpose of this relationship is to help you make some changes in your behavior. This cannot happen if you don't practice [the prescribed activities] outside these interviews.

Assuming the client wants to change behavior, the helper then assists in establishing realistic goals:

- HELPER: You say you're concerned about your work not having any meaning for you anymore. What would you like to see happen in this situation?

Helper responsibility for shaping goals

Although the motivation to change must come from the client, the helper has a responsibility to assist the client in shaping goals in several ways. First, goals must be realistic and achievable for the client and the situation in which the client will have to function.

- EXAMPLE: Your client is five-foot three-inches tall, and his goal is to play first-string professional basketball. The odds are that this is an unrealistic goal (and therefore unachievable). The odds are that the client is too short to play professional basketball because practically nobody under six feet tall can realistically hope to make the playing team.

However, size or other limitations may not prevent one from becoming actively involved in activities *related to* an area. Sometimes people fail to look beyond the concerns of the moment, automatically dismissing options by thinking of only one particular aspect of a situation (e.g., playing professional sports), instead of looking at the larger picture. In fact, even though a specific goal may be unrealistic, there may be a variety of vocational and leisure goal-related options from which the individual can acquire much satisfaction, such as coaching Little League, working in an athletic organization or sporting goods store, or participating in one of many options associated with a particular sport. Perhaps a hobby, leisure interest, or personally rewarding activity could be expanded into part-time or full-time employment options (Liptak, 1990).

Caution: Never say never! Be aware that the odds or probability of certain persons achieving noble goals may be slight, but many individuals have achieved remarkable, nearly impossible goals despite tremendous odds. Your client may be one of those persons.

The helping relationship: Opportunities and options

The helper has a responsibility to assist clients in looking beyond the immediate concern or problem, to consider what things could be like. Too often, clients are self-limiting—that is, concerned only with resolving or getting past the immediate problem or crisis. Such individuals are continually fighting fires, moving from one crisis to another, without planning fire prevention. Sometimes clients are so enmeshed in the crisis of the moment that they are locked or blocked in and cannot see an obvious solution. The helper can assist the client in perceiving the possibilities beyond the current situation. The *imaging technique* is one way to catch a glimpse of what could be. The helper assists the client in establishing goals that not only will resolve the current problem but also create options for preventing similar, future problems. (See Chapters 15 and 20 for more detailed information on how to do this.)

Imaging techniques bring out options

One way of helping clients get out of the limited-thinking, no-option rut is to have them imagine or think about what they would really like to do if there were no limits (see Chapter 19). This technique has three simple steps (Figure 11.4):

FIGURE 11.4 Imaging the future. The helper assists the client to (1) fantasize the ideal situation in which anything is possible. From this, (2) general characteristics of the future fantasy image are identified. Finally, (3) specific, realistic options are selected that can result in the client achieving a number of the fantasy characteristics. Remember that some people will be able to achieve their fantasy goals so, if practical, the helper should talk about the probabilities of fantasy achievement for the client.

1. *Imagine or visualize the fantasy*. Encourage clients to clearly imagine or visualize what they would like to do in the future—as if in a fantasy world where they could do anything they wanted (the five-foot three-inch client *could* play professional basketball).

2. *Assess general aspects*. Ask clients to generalize the major satisfactions they would get from this fantasy if it were true (being well paid, receiving recognition from others, having status, being an important part of a team).

3. *Determine realistic options*. From generalizations (item 2) the helper assists clients to determine important aspects of the fantasy that provide realistic potential options to pursue. In the example of the five-foot three-inch tall client who wanted to play professional basketball, the helper might assist the client in considering career options such as television or radio sportscaster, sportswriter, coach or manager, owner/manager, or salesperson in a

sporting equipment store plus a host of leisure and recreational volunteer options such as coaching, playing, or refereeing basketball with local groups or participating in athletic organizations and support groups.

Imaging can be very effective. If J. W. Marriott had not been able to visualize something more than selling root beer and hot dogs, the multibillion dollar Marriott Corporation might still be a Washington, DC, Hot Shoppe. Similarly, a client who owned a musical instrument shop greatly expanded its potential by visualizing a home entertainment center and branching into radios, high-fidelity equipment, and audio-video laser disk equipment. Hospital administrators have successfully used the same technique. They have enlarged their image of the traditional hospital—a place to treat physically sick patients—to include chemical-dependency treatment centers and health and wellness centers with exercise, nutrition, and fitness personnel and equipment to keep people healthy.

Visual imagery works extremely well with certain individuals. Athletes learn to visualize themselves moving through specific events in a flawless manner. Looking at videotapes of Olympic Game events, one can clearly see some athletes mentally progressing through the sequence of steps toward the goal, even moving their bodies subtly and tensing muscles as they visualize flawless performance in upcoming events. An unemployed teacher who can only see the possibility of working with school-age children is bound to be frustrated. In contrast, the teacher who recognizes that many of the teaching and interpersonal skills are involved in dealing with people in a wide variety of positions has a multitude of options available.

Balance enthusiasm with reality

Be positive and encouraging to the client, but temper enthusiasm with a healthy dose of reality. Most goals worth achieving will be fulfilled because of the client's dedication or commitment to their success. The typical client would like a magic wand to be waved so that the goal could be achieved immediately. Because most helpers have yet to perfect this ability, it is best to ensure that the client knows the reality of the situation and accurately anticipates the time and effort that goal achievement will take. A realistic awareness of what is required helps the client avoid discouragement when obstacles arise (Hoppock, 1976).

Summary

1. When the client achieves goals of the helping process, changes from current behavior in combinations of thoughts, feelings, and actions will occur.
2. Goals must be positive and realistic for the client.
3. Five guidelines for stating goals are the following:

   a. Goals must be stated so that they can be achieved.
   b. Conditions of performance must be specified.
   c. Consequences of goal achievement must be understood.
   d. Larger goals must be broken down into parts.
   e. Each goal must be measurable in some way.
4. The TFA triad provides clues to the client's current behavior and directions for behavior change—continuing or stopping current behavior or learning new behavior.
5. The helper assists the client in resolving current concerns and in looking beyond the immediate situation to see opportunities.
6. Broad general goals or more complex goals may need to be broken down into parts and made more specific to facilitate achievement.

Practice

1. Rule off a piece of paper so that it looks like Table 11.1. Label this page "Analyzing Goals."
2. Find a person who agrees to help you practice establishing realistic goals.
3. Select a role-playing situation that follows these instructions. Before starting your mini-interviews, clarify specifics of the situation: age and gender of the client, setting of the interview, and any other

TABLE 11.1 Analyzing Goals

Client's Concern: ______		
Goals (and/or Subgoals) That Will Resolve the Client's Concern	**Potential Positive Consequences**	**Potential Negative Consequences**

aspects essential to getting started. Assume that you and your coached client have already been talking for about twenty-five minutes.

4. Prior to role playing, set up a behavior blueprint using the TFA triangle method. Write in three or four of the most probable thoughts, feelings, and actions for your client's problem situation. Then have a helping interview with the client about possible goals in light of these specifics.
5. Conduct a five-minute mini-interview for each role-playing situation on the list. Record your interaction or have another person observe the interview, especially noting aspects related to establishing goals. Either the helper or client can start the interview by summarizing the client's concern. Then, work to establish at least one realistic, achievable goal during the next five minutes. After about four minutes, summarize the client's goal(s), noting relevant priorities in time and sequencing.
6. Following each mini-interview, ask your coached client for feedback about the interaction. Then listen to the tape or talk to your observer and fill out one of the goal-analysis pages you constructed. When you have completed the goal-analysis form, answer the questions in the "Focus Checklist."
7. Try some mini-interviews in which you practice appropriately integrating the various techniques learned up to now. Use the Behavior Summary Checklist at the end of this chapter.

Role play

Assume that these interviews have been going on for about twenty-five minutes to give you practice establishing goals.

1. A high school senior whose heart is set on attending a very expensive college more than 1,000 miles from home has not yet decided on any career goals but would most likely lean toward something that would require a general arts and sciences background. Parents and the client are having constant arguments about this situation. The client's first statement during the interview was "I've always wanted to go to [name of college]. That's where my best friend goes. My parents just don't understand. If I can't go there, I won't go to college at all."

2. A college sophomore enjoys indoor work rather than outdoor work. This client enjoys working with people, has a high grade-point average, and is currently a math major. The client's first statement in this interview was, "I just can't make up my mind about a career. This is bothering me so much that I lie awake sometimes thinking about it."

3. A twenty-six-year-old (single) client's life goal is to be financially well off. Your client has a college degree in business management and is currently selling life insurance. Early in the interview the client said, "When I started

selling insurance, I thought it was going to be great. Now I realize it's not what I expected, and I'm really at a loss for what to do."

4. This client has been feeling "tired with no energy." Although there are not any medical problems of which the person is aware, he or she has not had a physical examination for several years. This person rarely gets any physical exercise, has a desk job in an office, and spends the rest of the time in sedentary activity. The client's first statement during the interview was, "I guess maybe I ought to be doing something to get more energy, but when I get home from work, I feel really pooped out and just watch TV."

5. This client, between thirty-five and fifty-five years of age, is unhappy with work but has no idea what would be a better job and feels as is if he or she is just going through the motions of living. The first statements the client made during the interview were, "Life just seems to be so boring. Nothing much interests me anymore."

Focus checklist

1. Write the goal and subgoals, if necessary, that will resolve the client's concern.
2. Is each goal stated in specific, positive terms that describe what the client will be doing in the future?
3. Rank the behavior focus of the goal changes that are to be made in the client's thoughts, feelings, and actions.
4. List positive and negative consequences associated with achieving the goal(s) on a form similar to Table 11.1.
5. Is the goal realistic for your competencies?
6. Is the goal appropriate for the situation in which the client is (or will be) involved?
7. Is a time specified for the client to achieve the goal?
8. Is referral desirable or essential to help the client?
9. Using the TFA behavior blueprint, are the goals such that they would resolve the client's major concerns that were outlined in the problem situation?

Summary checklist

How many of the following elements did you use, and in what ways did you use them during your more integrated mini-interviews?

1. Structuring the helping relationship
2. Focus on thoughts, feelings, actions, or TFA combinations
3. Listening for understanding and nonverbal aspects
4. Asking open questions
5. Asking closed questions
6. Systematic inquiry
7. Clarification

8. Reflection
9. Use of silence
10. Confrontation
11. Client concerns
12. Goals

Instructions

To assess your interview skills, make a worksheet similar to the Behavior Summary Checklist. In the *left column* on the checklist, make a transcript of key words and phrases or sentences used by you and your client during the mini-interview. Then, *for each* transcript response, put a check in columns that are relevant. You might check from one to several items for each response. Ignore columns with content area that has not been discussed.

When you have completed the Behavior Summary Checklist, review it carefully. Look for patterns where you emphasized or ignored content. Then write a brief summary of the mini-interview outlining your strengths, limitations, and suggestions for improving skills.

References and recommended reading

Blanchard, K., & Johnson, S. (1981). *The one-minute manager*. San Diego: Blanchard-Johnson.

Bruce, P. (1984). Continuum of counseling goals: A framework for differentiating counseling strategies. *Personnel and Guidance Journal, 62*, 259–263.

Caulfield, T. J. (1986). The process of counseling: A synthesis. *Counselor Education and Supervision, 25*, 306–312.

Childers, J. H., Jr. (1987). Goal setting in counseling: Steps, strategies, and roadblocks. *The School Counselor, 34*, 362–368.

Corey, G. (1991). *Theory and practice of counseling and psychotherapy* (4th ed.). Pacific Grove, CA: Brooks/Cole. This book presents an excellent, highly readable, basic overview of major approaches to the helping process in counseling.

Cormier, W. H., & Cormier, L. S. (1991). *Interviewing strategies for helpers: Fundamental skills and cognitive behavioral interventions* (3rd ed.). Pacific Grove, CA: Brooks/Cole. This exceptional book, in all respects, outlines interpersonal aspects of interviewing and counseling skills and provides detailed descriptions of many behavior-change strategies and cases. The authors' personal commentary and experiences make this book a must for one who is planning to enter the counseling profession.

Dixon, D. N. & Glover, J. A. (1984). *Counseling: A problem-solving approach*. New York: Wiley. This book contains a wide variety of problem-solving approaches of general and specific use.

Glasser, W. (1981). *Stations of the mind: New directions for reality therapy*. New York: Harper & Row. Glasser is the founder and developer of reality therapy.

Hoppock, R. (1976). *Occupational information* (4th ed.). New York: McGraw-Hill.

Hutchins, D. E., & Vogler, D. E. (1988). *TFA systems*. Blacksburg, VA: Author.

Krumboltz, J. D. (1988). The key to achievement: Learning to love learning. In G. R.

Summary Behavior Checklist

Key words and phrases of the Client (CL) and the Helper (H)

T F A

Focus or Emphasis	
T	
F	
A	
Structure	
Nonverbal aspect	
Open question	
Closed question	
Systematic inquiry	
Clarification	
Reflection	
Confrontation	
Use of silence	
Client's concern	
Goals	
Procedures	
Evaluation	
Termination	
Followup	
Strategies	

Waltz (Ed.), *Building strong school counseling programs* (pp. 1–39). Alexandria, VA: American Association for Counseling and Development.

Lazarus, R. S. (1984). On the primacy of cognition. *American Psychologist, 39*, 124–129.

Liptak, J. J., Jr. (1990). Development of the career exploration inventory (CEI) (doctoral dissertation, Virginia Polytechnic Institute and State University, 1990). *Dissertation Abstracts International, 51*, 1510A.

Maultsby, M. C. (1984). *Rational behavior therapy*. Englewood Cliffs, NJ: Prentice-Hall. Maultsby is the founder and developer of rational behavior therapy.

McDaniels, C. (1989). *The changing workplace: Career counseling strategies for the 1990s and beyond*. San Francisco: Jossey-Bass.

Naisbett, J., & Aburdene, P. (1990). *Megatrends 2000*. New York: Morrow.

O'Brien, R. (1979). *Marriott: The J. Willard Marriott story*. Salt Lake City: Deseret.

Sampson, J. P., Jr. & Krumboltz, J. D. (1991). Computer-assisted instruction: A missing link in counseling. *Journal of Counseling and Development, 69*, 395–397.

Stiles, W., Shapiro, D., & Firth-Cozens, J. (1988). Do sessions of different treatments have different impacts? *Journal of Counseling Psychology, 35*, 391–396.

Zajonc, R. B. (1984). On the primacy of affect. *American Psychologist, 39*, 117–123.

Designing Effective Procedures

12

Key points

1. Designing procedures refers to the process of helping the client take action to move from the current concern or problem to achieve goals.
2. It is best to have the client suggest procedures and strategies to use to achieve goals.
3. The TFA triangle and TFA triad assist the helper in suggesting what kinds of procedures are likely to be most valuable to the client.
4. The process of establishing procedures begins by building on client strengths and gradually supplementing more limited areas.
5. Positive and negative aspects of the client's behavior provide clues for specific behavior to be strengthened, stopped, or learned.
6. Procedures need to be designed in a sequential way so that each step can be successfully achieved, leading to the next step.
7. If a given task is too large, it needs to be broken down into smaller parts.

LET'S briefly review the problem-solving process. First, the helper works with the client to assess and identify relevant behavior (thoughts, feelings, and actions) in the problem situation using the TFA triangle as a guide (Chapter 10). This assessment results in a descriptive behavior blueprint of the client's thoughts, feelings, and actions. Second, the helper and client establish realistic, achievable goals that will resolve the problem (Chapter 11). Third, with the goal(s) clearly identified, they design procedures that will assist the client in changing behavior. Procedures can be viewed as steps leading from the client's concerns to goal achievement (Figure 12.1). Thus, in a specified situation, the problem-solving process becomes

1. Assessing and identifying thoughts, feelings, and actions associated with the client's concerns.
2. Establishing realistic goals.
3. Designing procedures that lead to goal achievement.

Planning and prioritizing steps

When helping the client consider *how* to reach the goal, each step is planned so that the client is likely to successfully complete it. Thus, procedures are constructed to provide the client with the basis for moving effectively from one step to the next.

With each success, client motivation tends to increase as the goal is approached. Picture the process as a set of stairs in which each step is a move toward the goal. Each step is only high enough, or difficult enough, so that the client makes an effort to accomplish the task but has a good chance of achieving it.

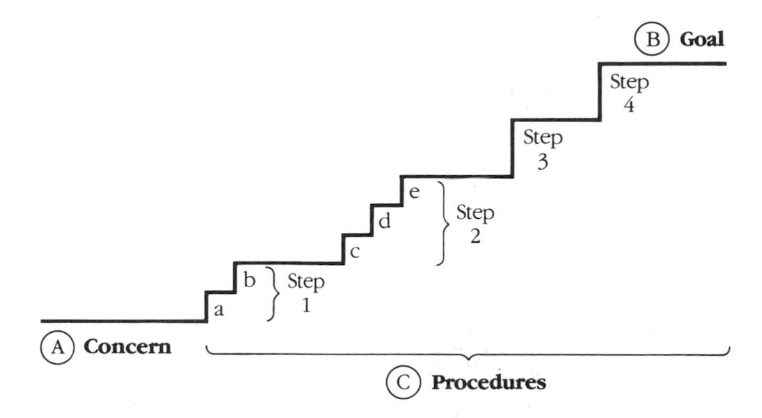

FIGURE 12.1 Designing effective procedures. The process of moving from (A) concern to (B) goal involves careful planning of (C) procedures. Steps 1 and 2 indicate how some procedures need to be broken into smaller steps (a and b, and c, d, e) in order to help the client complete them successfully. Note that the client must perform the substeps of steps 1 and 2 in order from a through e to complete these two difficult steps. (If the client were capable of accomplishing substep b before substep a, the step wouldn't have to be broken into substeps.)

Using the TFA triangle and triad

Because many aspects of the helping techniques have been used up to now, the helper knows a great deal about the client's thoughts, feelings, and actions in the problem situation. This knowledge and understanding of the client's behavior is absolutely essential because it is on this foundation that effective procedures can be designed to help the client change behavior. In the change process, the TFA triangle and triad are extremely useful in thinking about *how* to help the client *prioritize* procedures for change.

How do techniques of change relate to an individual client?

Although this is not a theory book, it is worthwhile to look briefly at the relationship between selected theories and associated techniques of procedures. Many procedures used to assist the client can be linked to major theories of counseling and psychotherapy. *Thinking procedures* are most frequently linked to cognitive, rational theories; *feeling procedures* are linked to affective, existential, and humanistic theories; and *acting procedures* are linked to behavioral theories. Of course, practically none of these are "pure." In referring to Hutchins' TFA model, Ellis (1980) wrote that nearly all theories and techniques combine aspects of the others. We have found it is practical and effective to conceptualize procedures in terms of how they tend to *emphasize* thinking, feeling, and acting dimensions of behavior.

Part of the helper's challenge is to *adapt* change procedures and strategies to the thinking, feeling, and acting orientation of the client (Hutchins, 1984). Figure 12.2 is a general classification of theories, writers and developers, and selected techniques that relate to the TFA model. *Thinking* relates to rational, logical, cognitive theoretical approaches as espoused by such persons as Ellis, Maultsby, Beck, and Burns (see Chapter 1 references). Techniques that apply to clients with a thinking orientation include those that appeal to the intellect: reading, lecture, tests, assessment, conceptual writing, facts, and information.

Feeling relates to affective, client- or person-centered, existential, gestalt, and humanistic theoretical approaches as espoused by such persons as Rogers, Perls, Patterson, and May. Techniques that apply to clients with a feeling orientation include those that appeal to emotions: affect, experiential, sharing and group interaction, encounter, gestalt, music, movement, and role play.

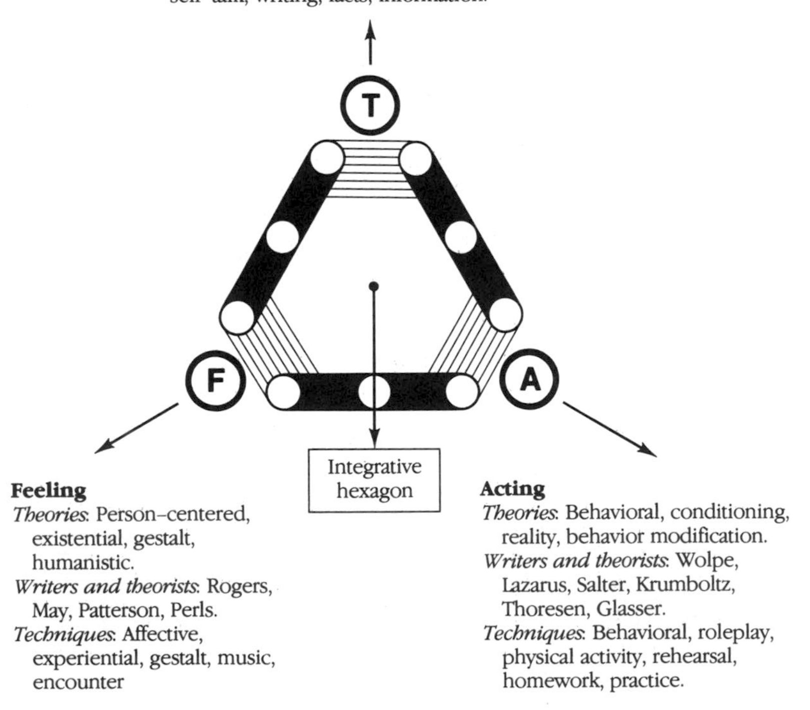

FIGURE 12.2 The TFA triangle illustrates selected major theories, writers and theorists, and techniques associated with thinking, feeling, and acting areas. The integrative hexagon is where thoughts, feelings, and actions are blended in various combinations.

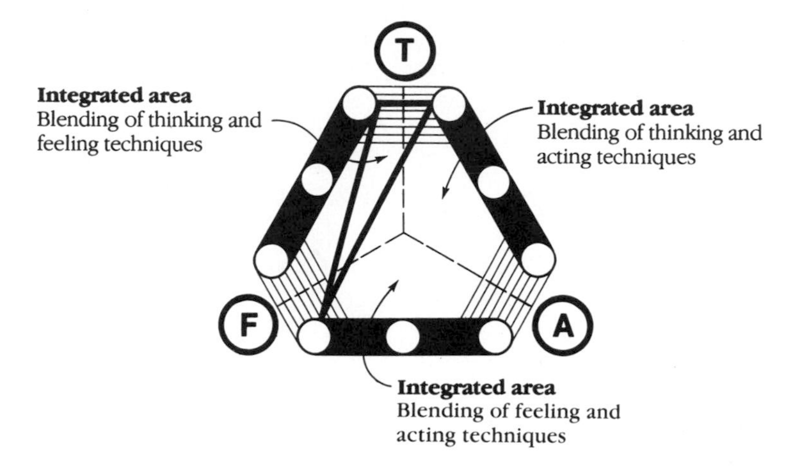

FIGURE 12.3 The TFA triangle illustrates a thinking–feeling triad and the three integrated areas in which various techniques are blended. Shaded areas at each vertex illustrate where predominant thinking, feeling, or acting techniques might be best employed.

Acting relates to behavioral, behavior modification, reciprocal inhibition, conditioning, and reality theories as espoused by such persons as Wolpe, Lazarus, Salter, Krumboltz, Thoresen, and Glasser. Techniques that apply to clients with an action orientation include those that have the client doing things: simulation, behavior rehearsal, manipulation, investigation, practice, physical activity, and involvement in a variety of activities both in and out of the helping interview.

As a working model, our experience indicates that the closer to one of the thinking, feeling, or acting vertices of the TFA triangle, the more effective it is to use techniques and strategies that emphasize characteristics as outlined above. Thus, if the client's TFA triad is dominant in one or more areas, use techniques associated with that approach. Figure 12.3 suggests that cognitive techniques may be most helpful because of dominance of the triad in the thinking area. This approach would be bolstered with techniques in the affective–feeling area as well as recognizing the client's thoughts. Note that the client is not likely to take action in this situation.

Therefore, the client is most likely to respond to ideas (T) that can be related to feelings (F) that reflect caring and concern; and *if* these appeal to both thoughts and feelings, the probability that the client will take appropriate action will likely increase. In the center of the triangle is the *integrative hexagon* (Figure 12.2). When the client's triad is in the hexagon, the helper is able to *combine, integrate,* or *blend* approaches. Most helping approaches actually tend to overlap and blend these, as in the case of a humanistic behaviorist who uses behavioral approaches in a sensitive and caring way, or a cognitive behaviorist (such as Meichenbaum) who blends the thinking and acting approaches, or the rational emotive therapist who emphasizes thinking

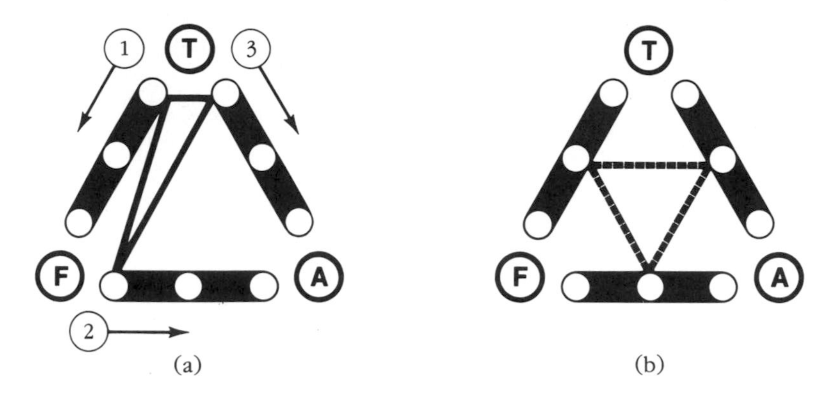

FIGURE 12.4 (a) This triad illustrates the sequence that takes place in working with a client with a thinking–feeling orientation. The numbers 1, 2, and 3 show the sequence of the helper's probes, which build on client strengths. (b) This triad indicates a midpoint orientation, which is the initial problem-solving goal. The midpoint triad is most desirable because the client can perceive all thoughts, feelings, and actions that affect any specified situation.

and feeling techniques while directing the client to engage in action strategies and homework. Figure 12.2 outlines theoretical orientations and sample techniques associated with the TFA system.

Note that the integrative hexagon is further divided into three sections (Figure 12.3). Each section suggests an integration, or blending, of techniques and strategies that may be most helpful for a particular client. For example, on the left, thinking and feeling are blended; on the bottom, feeling and action; and on the right, thinking and action. An awareness of these concepts can be useful in addressing the client's strengths in the problem situation. Using the TFA triangle and the client's TFA triad, the helper can select rationally and systematically, building on the client's strengths, theories, techniques, and strategies for behavior-change procedures. The helper's alertness to the client's ways of responding to people, data, and things increases the helper's ability to attend to all the client's physical senses and the thoughts, feelings, and actions.

Practical application of TFA procedures

How do the preceding concepts work for a particular client? To apply the TFA procedures in choosing strategies to assist in the change process, let's follow the client we have been using through Chapters 9–11. Recall that this problem situation has a thinking–feeling triad as Figure 12.4 shows. The problem statement is "I'm afraid to talk with and interact with others in a social setting where I don't know other people." The *goal* is "to learn to interact effectively with others I don't know in a social setting."

The greatest part of the triad falls into the thinking area, with a somewhat lesser amount in the feeling area. This pattern suggests a high probability that the client is most likely to respond to thinking (cognitive, rational, logical) approaches and feeling (emotional, affective, caring) approaches that are part of the person's behavior in the problem situation. Furthermore, because a large portion of the client's triad falls into the integrative hexagon, a blending of thinking and feeling procedures is most likely to motivate the client to change behavior. Thus, what the triad suggests is *how the helper might best adapt to the client's behavior orientation*.

Note two things. First, the helper can adapt to the client's mode of behavior. If this T–F triad is the client's typical behavior pattern, the adaptation is likely to result in the client "feeling good" about the *relationship* with the helper. Second, as Figure 12.4 illustrates, *it is this particular triad pattern* (specific thoughts and feelings, as well as a relative lack of action) *that is associated with the client's ineffectiveness* in the situation. Therefore, the *behavior pattern,* as reflected by the TFA triad and associated behavior, may need to be changed if the client is to function effectively.

Therefore, behavior that is different from the client's current TFA triad is exactly what is needed to help the client shift from a comfortable (well-practiced) behavior to more effective behavior. To illustrate, this concept is seen in various techniques related to one's thoughts, feelings, and actions:

T: Paradoxical intention gets the client to think the exact opposite of what would seem reasonable.
F: Top dog/underdog is a gestalt technique that enables the client to express feelings as well as thoughts about others.
A: Actually doing what one fears (e.g., playing a guitar and singing in front of people in Central Park) is one way of reducing anxiety about performing in front of others.[1]

Sequencing procedures

The helper can use the following guidelines to consider the appropriateness of procedures for a specific client with a particular TFA triad orientation.

- GENERAL PRINCIPLE: The midpoint TFA triad is the best *initial goal* for effective problem solving.

The reason? The midpoint is the best position for the client to see all options. The client considers all thoughts, ideas, and options; weighs feelings and emotions against alternative ideas; and projects consequences of potential actions against ideas and feelings in a continuing, mutually interactive feedback process (Figure 12.4b).

To illustrate this principle, let's continue with the client we have followed throughout the problem-solving stage of the helping process. At the start of the helping process, when the client's TFA assessment is made, the

[1] Thanks to Paul Little.

client has a strongly T–F triad (Figure 12.4). First, the greatest probability of this client's behavior pattern is in the thinking area where the person continuously, perhaps ponderously, thinks about the situation. Second, the client's pattern shows feelings and emotions, which reciprocally affect both what and how the person thinks and so on in a continuing cyclic process. The mutually established goal is to enable the client to become more socially interactive (Chapter 11). To do this, first *build on* the client's strengths (as seen in the TFA triad in Figure 12.4) and then *supplement* the limited areas where the triad is not present (Hutchins & Vogler, 1988). Note that a client's "strengths" are defined here as the *behavior that is most likely to occur*. A positive or negative connotation is not necessary. The helper builds on client strengths as a way of capitalizing on each client's individual behavior pattern in the problem situation. For the client in Figure 12.4, the specific procedural steps begin with the client's strengths in thoughts and feelings, gradually moving to shift behavior toward a midpoint of more action. The helper and client work together to determine specific kinds of procedures so that the client could realistically accomplish the sequential steps toward the eventual goal.

Break down larger steps into manageable parts

Some steps are so large that they must be broken down into smaller parts. For example, a goal of interacting with others in a social situation is desirable (and achievable) but may have several components that can be learned on a step-by-step basis. Procedures might include:

1. Learning to give appropriate eye contact.
2. Learning to introduce self to others without seeming overbearing or prying.
3. Giving appropriate nonverbal behavior.
4. Asking questions and attending to what others say and do.
5. Learning to start and keep a conversation going.

When the client achieves these five *procedures*, the client also fulfills the overall goal of learning to interact with others (Figure 12.5).

Positive and negative aspects of behavior

The second set of considerations involves positive and negative aspects of the client's behavior. In assessing and interpreting behavior up to this point the helper has determined the client's *behavior blueprint* for the problem situation. Recall from Chapter 10 that the behavior blueprint is a description of specific thinking, feeling, and acting dimensions of behavior that may contain both positive and negative aspects (Figure 10.2). These elements clearly point out positive and negative aspects of thinking, feeling, and acting related to the problem. Positive aspects can be used to help in the change process. Negative

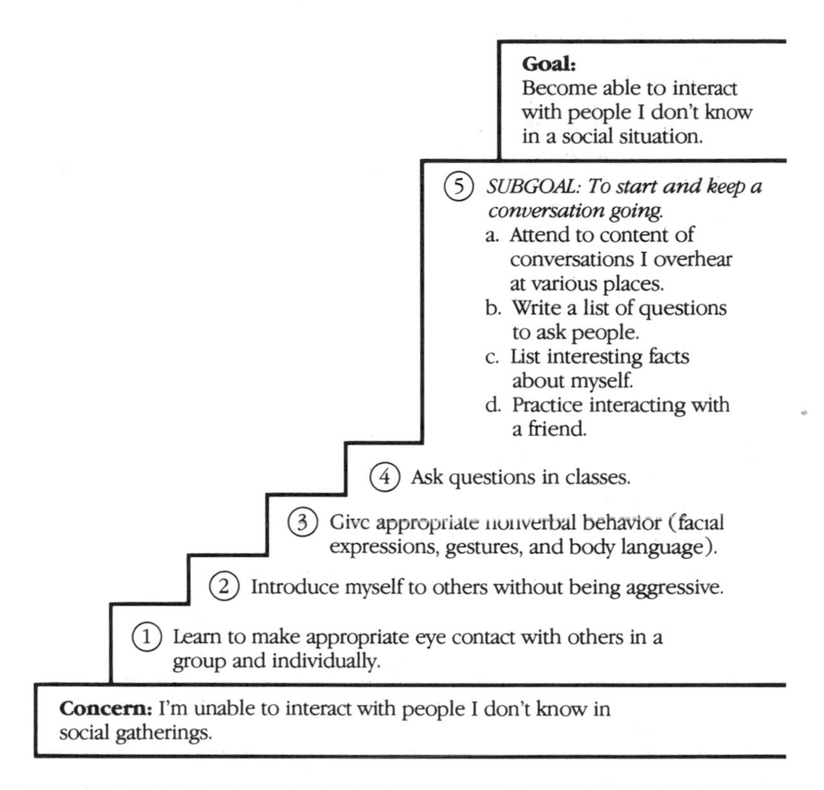

FIGURE 12.5 Breaking tasks into manageable parts. Five steps were planned to assist the client in achieving the overall goal. The client accomplished the first four steps relatively easily but had trouble with step 5. Therefore, step 5 became a subgoal and was broken down into four smaller parts (a–d). When each part had been completed, the client could interact easily with people in a social situation.

aspects must be changed if there is to be progress toward achieving helping goals.

Examine your assumptions about what the client can do

Too often, a well-meaning helper makes assumptions about what the client can or cannot do and skips over essential beginning steps that are a prerequisite to other behaviors. Examples include telling another, "Don't worry," "Be more assertive," or "Get involved with others." This is equivalent to telling the client who does not interact with others to "go out and make friends." If such a client could do this, there would be no need for a helper. It is precisely because the client cannot do these things alone that a helper is needed to provide concrete assistance.

The key question for procedures is, *What can the client do successfully now?* If the client can successfully complete the step, fine. Move on to the next step. But if the client is unable to successfully complete a given task, it becomes a *subgoal,* which is then broken down into smaller parts. More explicit steps are then set up to assist the client in achieving the subgoal. Figure 12.5, step 5, illustrates this process.

Stop, strengthen, and learn behavior

Another way of looking at procedures is to look at the client's behavior blueprint (Figure 10.2) to review specific aspects that are positive and negative. The helper assists the client in looking at behavior that must be stopped or reduced, behavior that can be strengthened or increased, and new behavior that must be learned. Sometimes it is helpful to plot these into a chart and list specific thoughts, feelings, and actions that need to be stopped, continued, or learned. Table 12.1 illustrates this process. Then, the helper can design specific kinds of activities to help the client accomplish these specific behaviors.

Cues that indicate difficulties

As the client talks about specific steps that might be taken, the helper must be especially alert for subtle cues that suggest difficulties or problems for the client. These cues may be found in both verbal and nonverbal behavior.

In the five-step procedure, the goal is to interact more effectively, in a

TABLE 12.1 Three Main Ways of Changing Behavior

Aspects of Client Behavior	Stop or Reduce this Behavior	Strengthen or Continue This Behavior	Learn New Behavior
Thoughts	Negative self-talk such as "I can't possibly talk with others."		Positive self-talk such as "What I say can be interesting to other people."
Feelings	"I get embarrassed in the presence of others."	Emphasize client's positive feelings about self and abilities.	
Actions	"I stand off by myself when I don't know other people."	Capitalize on what the client does well, like music, skiing, swimming, reading, and other interests.	Go where others are. Practice asking questions about interests of others.

(1) Stop or reduce a negative behavior, (2) strengthen or continue a current positive behavior, and (3) learn a new behavior. Each activity may be associated primarily with thoughts, feelings, or actions, which are the major dimensions of behavior that need to be changed by the client.

social situation, with persons who are not known (Chapter 11). Assuming the client's commitment to change behavior, what tasks does the helper "assign" to the client? This depends on the individual and circumstances. One of the *best* ways of knowing what a person can do and how well is to have the client *role-play* a typical situation with the helper during the interview. The helper can observe the nonverbal hesitations, the verbal stammering as the client reaches for appropriate words to use, the pauses and silence that suggest the client does not know what to do, and so on. Thus, in looking at Figure 12.5, steps 1 and 2 (eye contact and introducing self), might be role-played or rehearsed in the helping interview. Following the role play or rehearsal and discussion of how to improve it, the client is given homework in which specific tasks are practiced until they become habits.

But look carefully at step 2. If the client is given homework or practice to introduce self to others outside the helping interview, be aware that the client does *not* currently have skills outlined in steps 4 and 5. Without knowing *what* to say (after saying, "Hi . . . My name is __________"), the client is primed for *failure*. The client will feel embarrassed and ill-at-ease, leading to more self-deprecating thoughts ("I didn't *think* this would work"). All this leads to reducing the probability of acting in an appropriate manner (talking to others) in the future. Therefore, work with the client to do things only where there will be a high probability of success. If you or the client are not convinced a task can be done successfully, this is a good indication that more supervised practice is needed in the helping interview, before the client takes on the task as homework. In this case, steps 1–3 should be practiced in a *safe,* supportive, encouraging environment such as the helping interview or with friends who agree to help—a peer support group or group counseling. Following successful experiences and practice with positive, reinforcing consequences, the client is then able to move out into the "real world." Attention to specific aspects of how the client thinks, feels, and acts can alert the helper to potential problems.

1. *How the client thinks.* By carefully attending to not only what the client says but also how it is said (enthusiastically vs. hesitatingly), the helper can detect problems the client may have in understanding aspects of what, where, how, and when things might be done.
2. *How the client feels.* Is the *energy level* of the client high, moderate, or low?[2] By noting verbal and nonverbal hesitations, frowns, halting voice, silence, and body language, the helper can be alert to things that don't seem to "feel right" to the client.
3. *How the client acts.* By being aware of how the client talks, body posture, gestures, and facial expression, the helper can further detect potential concerns or problems related to doing whatever is necessary to change behavior.

Although a good general objective is to set up successful experiences, the odds indicate some failure experiences. The helper has a responsibility to

[2]Thanks to Maryam Sobhany.

help the client recognize that there will be *some* (perhaps many) failure experiences—especially when the client does not have control over others. The client will contact persons who are mean, rude, or obnoxious to *most* people, a supervisor who does not care about others, or a person with whom the most likable person in the world could not interact. If the client is prepared for such experiences, the helper *increases* the client's ability to deal effectively with the situation, especially if such events have been role-played and processed in the helping sessions.

The importance of homework

It is highly desirable to have the client work on relevant personal behavior outside the interview setting—to do homework. For several reasons, assigning specific homework can greatly facilitate the client's progress toward achieving goals. For example, you might have the client read specific articles, books, or specific chapters on a relevant topic; keep track of specific actions; try out new ways of doing things; or practice new skills or ways of interacting with others or engaging in new behavior patterns. *Always have the client do something related to the concern or goal*—find out more information, talk to people, keep a log or journal related to specific thoughts, feelings, and actions, experiment with new behavior, and so on. Homework helps in many ways:

1. Homework keeps the client working on relevant behavior between interviews. Although some behavior change happens in the helping interview, the goal is to change behavior outside the interview.
2. The client can return to the next interview with specific information, detailed personal observations, descriptions of what happened, and the like.
3. Completing homework indicates that the activities are appropriate and the client is motivated to change behavior.
4. Not completing activities provides evidence that may be related to the client's motivation or the appropriateness of activities, as well as a discussion of other aspects that may not have been apparent.
5. Evidence that the client does or does not complete homework provides a means of evaluating and modifying activities to make them more effective (see Chapter 13 on evaluation).
6. Homework allows the client to practice relevant behavior in a natural environment or real-life setting and in the presence of persons with whom one normally associates.[3]
7. Homework helps give the client a greater sense of responsibility for and control over personal behavior (see Chapter 17 on charting behavior).

[3]Thanks to Marjie Flanigan.

If the client understands what is to be done (along with other specifics of procedures such as how to do it and when, where, and under what conditions), the critical factor becomes whether it can be done successfully. The helper's objective now is to see that procedures are appropriate and realistic for the client's situation. If a client expects to be successful in performing the tasks necessary to achieve personal goals, the client will most likely experience increased motivation and personal satisfaction that result from accomplishing goals.

Caution: Be careful not to overwhelm the client with too many activities! It is better that the client can do a few things successfully than to have many activities that result only in failure or partial task completion. Successful completion of procedures leads the client to increased confidence in doing tasks effectively and helps provide the increased motivation essential to progress toward major, longer-term goals. Giving encouragement and support to the client who tries out new behavior is desirable. However, the greatest satisfaction should occur naturally to the client, by experiencing the pleasure of effective personal behavior.

Help the client identify and use resources

The helper can be of tremendous assistance in identifying resources that can help the client take positive steps toward the goal. In many ways, the helper acts as an information source by first helping the client consider options and then making suggestions about how the many different options might be used. Resources are grouped into three broad categories:

1. *People*: This category includes family, friends, peers, neighbors, and associations with others in various kinds of social, health and recreational, community, agency, government, educational, volunteer, and religious organizations at local, state, national, and international levels.
2. *Data*: This category includes programs; workshops; television; audio–video computer tapes and disks; newspapers, magazines, brochures, books, and journals; telephone hot-line and crisis information; written records and information; tests and appraisal instruments of all kinds including aptitude, attitude, interest, ability, and intelligence tests; mental and emotional measures; personal orientation inventories, résumés, and work samples.
3. *Things*: This category includes special facilities and equipment including those for personal, social, health, and physical assistance (wheelchairs, crutches, biofeedback hardware and software, exercise and hospital equipment for home use); special items for use by physically or mentally disabled; facilities and shelters for battered spouses and children; detoxification centers; jail; and emergency treatment and longer-term care centers.

The client does not have to do it all alone. Many existing resources (people, data, things) are available for the client and the situation. The helper should be ready to assist as a resource person by suggesting options and appropriate referral sources in the behavior-change process.

Summary

1. Procedures are designed so that successful completion of one step leads directly to the next step in a sequence as the client moves from the concern to achieving goals (Figure 12.1).
2. The TFA triangle and triad can be used to help the client make changes because the helper builds on client strengths and then supplements these with new behavior.
3. It is better to have the client engage in tasks that can be successfully completed—and increase motivation—than to overwhelm the person with too much to do, which results in failure.
4. Larger steps become subgoals that are broken down into smaller parts to help the client be successful in completing them (Figure 12.5).
5. A client is often in the best position to know the kinds of things that can change behavior.
6. The helper reinforces the client in taking positive steps in the behavior-change process.
7. In designing effective procedures, the helper builds on the client's strengths and then supplements these with procedures related to more limited areas.
8. Procedures and techniques to emphasize can be determined by looking at the client's triad to see what areas of the TFA triangle it covers.
9. The helper tries to anticipate potential difficulties that the client may experience in accomplishing each step in the process.
10. Positive and negative aspects of behavior provide clues for behavior that needs to be stopped, strengthened, or learned in terms of the client's thoughts, feelings, and actions (Table 12.1).
11. Verbal and nonverbal aspects of behavior can provide cues that suggest when the client may not understand things (thoughts); may feel uncomfortable, anxious, or uncertain about things (feelings); or doesn't know what to do (actions) in the change process. The helper who is alert to these subtle messages can greatly assist the client by designing experiences that increase the probability of changing behavior.
12. Homework, or assignments and activities outside the interview, gives clients a greater sense of responsibility and control over their own behavior; it also provides valuable, specific information on needed modifications that can increase the effectiveness of procedures.

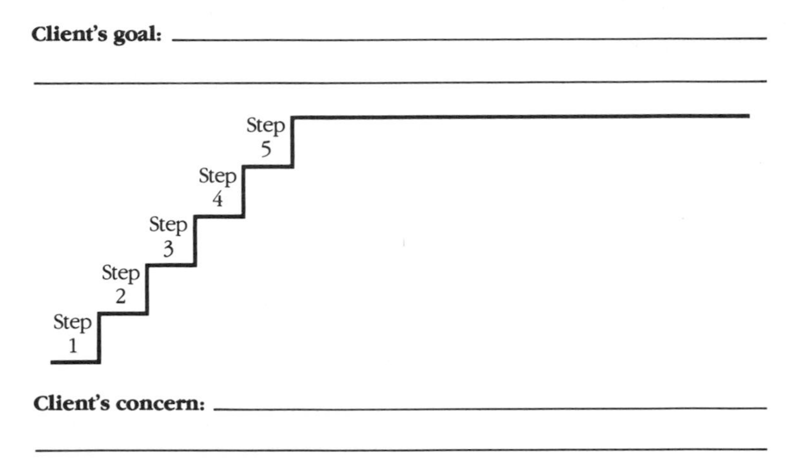

FIGURE 12.6 Designing effective procedures. List several (three to five) procedures or steps that will help the client in moving from the concern or problem to the goal. If there are steps the client may have trouble achieving, break them down into smaller parts to help ensure that they can be accomplished.

13. Resources to help the client achieve goals include people, data, and things, which are available at a personal, community, state, and national level.

Practice

1. Draw a diagram or copy Figure 12.6.
2. Find a person who agrees to help you learn how to design effective procedures.
3. Select one role-playing situation from those that follow. Before starting your interview, clarify specifics of the situation: age and gender of the client, setting of the interview, and other details essential to getting started. Assume that the interview with the client has been going on for about thirty minutes. Write the client's concern and major goal on the sheet (Figure 12.6) before starting this mini-interview. Record your interaction or have an observer take notes, especially on aspects of the interview that relate to designing procedures.
4. Conduct a five- to ten-minute mini-interview for each role-playing situation. During the last minute, either you or the client should summarize steps the client will be taking to work toward the goal.
5. Following each mini-interview, ask your coached client for feedback about the interaction. Then listen to the tape or talk to your observer and fill out one of the procedure-designing pages you made (Figure

12.6). Then answer questions in the "Focus Checklist," which follows the role-playing situations.

6. Finally, try some mini-interviews in which you practice appropriately integrating the various techniques learned up to now.

Role play

Assume the interview has been going on for about thirty minutes.

1. Your client, who has recently moved to this area, has made no friends yet and is unhappy and lonely. You and the client have established that the goal is to do whatever is necessary to make friends in this new location.
2. The client has trouble asking questions and expressing opinions and generally does not become involved in activities. The goal is to learn to interact socially with others the client does not know. Assume the client is motivated to do this but feels very self-conscious about being involved with others.
3. The client is concerned about being in financial difficulties. On the first of the month, both the client and spouse feel close to panic because they know there is not enough money to pay the bills and live in the manner to which they are accustomed. The goal is to develop a plan for handling and budgeting money.
4. The client has been unhappy at work for several months and has come to you because "going to work is really a pain." The goal is to help the client explore more satisfying employment. (You'll need to agree on what this person has been doing in previous work.)

Focus checklist

1. Are procedures designed in a sequential way beginning with steps 1, 2, 3, and so on?
2. Will the procedures resolve the client's concern or problem?
3. What verbal and nonverbal behavior indicates the client might have difficulty in doing what is necessary?
4. If there were potential problems, how did you handle these?
5. Are more difficult steps broken down into smaller parts?
6. Did you first try to get ideas from the client about what steps could be taken to resolve the problem?
7. Did you reinforce the client for developing personal ideas about how to proceed?
8. Did you provide needed information and offer suggestions appropriately (without giving advice)?
9. Did you assign homework to help the client practice specific behavior outside the interview setting?
10. Are tasks realistic for the client and situation?

11. Can the client be expected to successfully complete activities prior to the next interview?
12. Are all appropriate resources used in planning how to make the desired changes (including people, data, and things)?
13. Did helping procedures and strategies build on the client's greatest strengths?
14. Are procedures likely to overwhelm the client or are they realistic?
15. If necessary, did you refer the client to appropriate professionals?

Summary checklist

How many of the following elements did you use, and in what ways did you use them during your more integrated mini-interviews?

1. Structuring the helping relationship
2. Focus on thoughts, feelings, actions, or TFA combinations
3. Listening for understanding and nonverbal aspects
4. Asking open questions
5. Asking closed questions
6. Systematic inquiry
7. Clarification
8. Reflection
9. Use of silence
10. Confrontation
11. Client concerns
12. Goals
13. Procedures

Instructions

To assess your interview skills, make a worksheet similar to the Behavior Summary Checklist. In the *left column* on the checklist, make a transcript of key words and phrases or sentences used by you and your client during the mini-interview. Then, *for each* transcript response, put a check in columns that are relevant. You might check from one to several items for each response. Ignore columns with content area that has not been discussed.

When you have completed the Behavior Summary Checklist, review it carefully. Look for patterns where you emphasized or ignored content. Then write a brief summary of the mini-interview outlining your strengths, limitations, and suggestions for improving skills.

References and recommended readings

Cormier, W. H., & Cormier, L. S. (1991). *Interviewing strategies for helpers* (3rd ed.). Pacific Grove, CA: Brooks/Cole.

Summary Behavior Checklist

Key words and phrases of the Client (CL) and the Helper (H)

T F A

Focus or Emphasis: T	
Focus or Emphasis: F	
Focus or Emphasis: A	
Structure	
Nonverbal aspect	
Open question	
Closed question	
Systematic inquiry	
Clarification	
Reflection	
Confrontation	
Use of silence	
Client's concern	
Goals	
Procedures	
Evaluation	
Termination	
Followup	
Strategies	

Ellis, A. (1980). Rational-emotive therapy. In R. J. Corsini & D. Wedding (Eds.), *Current psychotherapies* (4th ed.) (pp. 197–238). Itasca, IL: Peacock.

Glasser, W. (1965). *Reality therapy: A new approach to psychiatry*. New York: Harper & Row.

Hutchins, D. E. (1984). Improving the counseling relationship. *Personnel and Guidance Journal, 62*, 572–575.

Hutchins, D. E., & Vogler, D. E. (1988). *TFA systems*. Blacksburg, VA: Author.

Ivey, A. E. (1983). *Intentional interviewing and counseling*. Pacific Grove, CA: Brooks/Cole.

Lee, C. C., & Richardson, B. L. (Eds.) (1991). *Multicultural issues in counseling: New approaches to diversity*. Alexandria, VA: American Association for Counseling and Development. This is an excellent and timely book that includes experiences of 17 experts in the area of multicultural counseling including Native Americans, African Americans, Asian Americans, Latin Americans, Southeast Asians, and Arab Americans.

Lee, C. C. (1991). Empowerment in counseling: A multicultural perspective. *Journal of Counseling and Development, 69*, 229.

May, R. (1983). *The discovery of being: Writings in existential psychology*. New York: Norton.

McDaniels, C. (1989). *The changing workplace: Career counseling strategies for the 1990s and beyond*. San Francisco: Jossey-Bass.

Vontress, C. E. (1981). Racial and ethnical barriers in counseling. In P. Pedersen, J. Draguns, W. Lonner, & J. Trimble (Eds.), *Counseling across cultures: Revised and expanded edition* (pp. 87–107). Honolulu: University of Hawaii Press.

13 Evaluation, Termination, and Follow-Up

Key points

1. Evaluation is the act of making judgments to determine the worth or merit of the helping process.
2. Formative evaluation starts with the initial interview and continues throughout the helping process.
3. Summative evaluation is conducted at the end of the helping process.
4. Evaluation includes
   a. The client's behavior in the problem situation.
   b. The helping process and how it relates to the client.
5. Changes in the client's TFA triad and different behavioral descriptions (the behavior blueprint) are useful in evaluating client progress.
6. Termination occurs when
   a. Goals have been achieved.
   b. The client's motivation or situation has changed.
   c. Progress is not being made.
7. In the termination process, the helper assists the client in using the problem-solving process to increase independence in other areas of life.
8. The follow-up process may involve monitoring progress, reinforcing gains and providing support, correcting minor problems, assessing changes, compiling data, and evaluating further the effectiveness of the helping process.

EVALUATION is "the act of rendering judgments to determine value—worth and merit . . ." (Worthen & Sanders, 1987, p. 24). Evaluation implies standards, or criteria, against which people and procedures are compared. In thinking about evaluation, Roffers, Cooper, and Sultanoff remind us that "the ultimate measure of training effectiveness is not skill acquisition, application, or retention but *whether* the trainees' *clients gain from their counseling experience*" [italics added] (1988, p. 388). This statement reminds us that, although this book focuses on what the helper does, this training is secondary to the primary focus: helping the client to change. Thus, the ultimate question is, Does the client gain or benefit from the helping process? Formative and summative evaluation are two ways of addressing this question.

In the helping relationship, a *formative evaluation* is conducted *throughout* the helping intervention. It starts when the client is first seen and continues until the relationship is terminated. To make these continuing evaluations, the helper assesses two major aspects: client behavior and the helping process.

Three questions summarize the assessment of the *client's behavior*:

1. What is the client's major concern or problem to be addressed in the helping relationship? (Chapter 10)
2. What is the client's goal? (Chapter 11)

3. What is the client's current behavior (at any point in the process)? (Chapters 2 and 12)

Answers to these questions are part of an assessment of the *client's behavior* and lead to evaluating its effectiveness. The next two questions relate to the helping process:

4. What theories, techniques, and strategies are being used in the helping process?
5. How effective are these theories, techniques, and strategies in helping the client to change behavior?

Evaluating the client's behavior and the helping process must occur simultaneously throughout the helping interaction. A continuing assessment of the client's behavior (1, 2, and 3) should lead the helper directly to (4) theories, techniques, and strategies that will have a maximum probability of helping the client. This, in turn, leads to evaluating (5) the effectiveness of theories, techniques, and strategies in the helping process. Question 5 cannot be answered without data relating to questions 1, 2, and 3. If the client's behavior does not change in positive ways, it may suggest that procedures used in the helping process need to be changed:

- HELPER: Last week you were going to set aside fifteen minutes each evening to talk quietly with your daughter. What happened?
- CLIENT: My work shift was changed to the night shift, and I found that I wasn't home at any time when we could talk.

In this example, the strategy was good but did not fit the client's new work situation. Because the situation changed, some other alternative must be devised. The same process can be applied to the helper's theoretical orientation and associated strategies:

- SITUATION: The helper has been using primarily action, or behavioral, strategies to help the client make changes but now sees resistance. The helper evaluates the action strategies as inappropriate and shifts to more of an affective, or feeling, orientation. In exploring the client's feelings and emotions, the helper finds that the client is reluctant to take action because of apprehension about consequences that had not been expressed.

As illustrated above, formative evaluation enables the helper to make adjustments in theoretical orientation and complementary techniques. Such an evaluation prompts a corrective adjustment in the helping process.

Evaluation and the client's thoughts, feelings, and actions

If you plot a client's behavior on the TFA triangle (Chapters 2 and 10), you have a graphic representation of the client's behavior in the specified problem situation. You also have an indication of needed goals as mutually agreed on by the client and helper (Chapter 11). As termination of the helping relationship approaches, the client's TFA triad, which indicates behavior, should be different from the original triad if there has been progress:

1. The TFA triad will be a different shape.
2. Specific behavior at each dimension of the TFA triangle will be different as seen in the behavior blueprint (Chapter 12) (e.g., more positive thoughts have replaced prior negative thinking, or more positive and assertive actions have supplanted self-defeating actions). Recall that the behavior blueprint details specific behavior descriptions on each side of the TFA triangle.

Such differences, along with client descriptions and helper observations, are indications of changes in the client's original behavior to the current behavior. Recall the comprehensive ABC model outlined in Chapter 2. *A* represents the situation or antecedent conditions associated with the client's concern. *B* illustrates the client's behavior (including thoughts, feelings, and actions) in the problem situation, and *C* includes the consequences of the client's behavior in the specified situation. In assessing the impact of the helping relationship, at least *B* and *C* will change, and in some cases the situation (*A*) will also change, especially if a physical move occurs to change the physical setting.

As an example, assume that the client was seen because of angry outbursts in which the client physically abused his spouse. Figure 13.1 illustrates a TFA triad that is strongly feeling and action-oriented in the problem situation. The assessment phase of the problem solving process established, in detail, the client's specific behavior. The primary goal was to stop abuse. Procedures were designed that built on feeling and acting strengths in the client's TFA triad: to change the client's behavior from physical expression of emotion (aggressive actions following feelings of anger), to become aware of relevant thinking, feeling, and acting cues, and to find positive means of venting emotions through constructive actions. Figure 13.b is an example of what one client's TFA triad looks like after an effective intervention by a trained helper. Notice the specific behavior changes illustrated by the shape of the client's TFA triad and behavior blueprint (see Chapter 12 for review).

The TFA triad has changed from one that is a very feeling–acting pattern (Figure 13.1a) to a midpoint pattern in which the client thinks about alternatives and chooses more positive options in the specific situation before violence occurs (Figure 13.1b). Also, as indicated by descriptive words at each point on the triangles, descriptions of the client's behavior changed from emotionality (F) without restraint (A) to showing much more thinking before taking action. This kind of behavior change and resulting differences in TFA triads has been clearly demonstrated in working with a variety of problem behaviors including domestic violence (Clow, Hutchins, & Vogler, 1990), adult victims of incest (Tieman, 1991), and pregnant teens (Bundy, 1991).

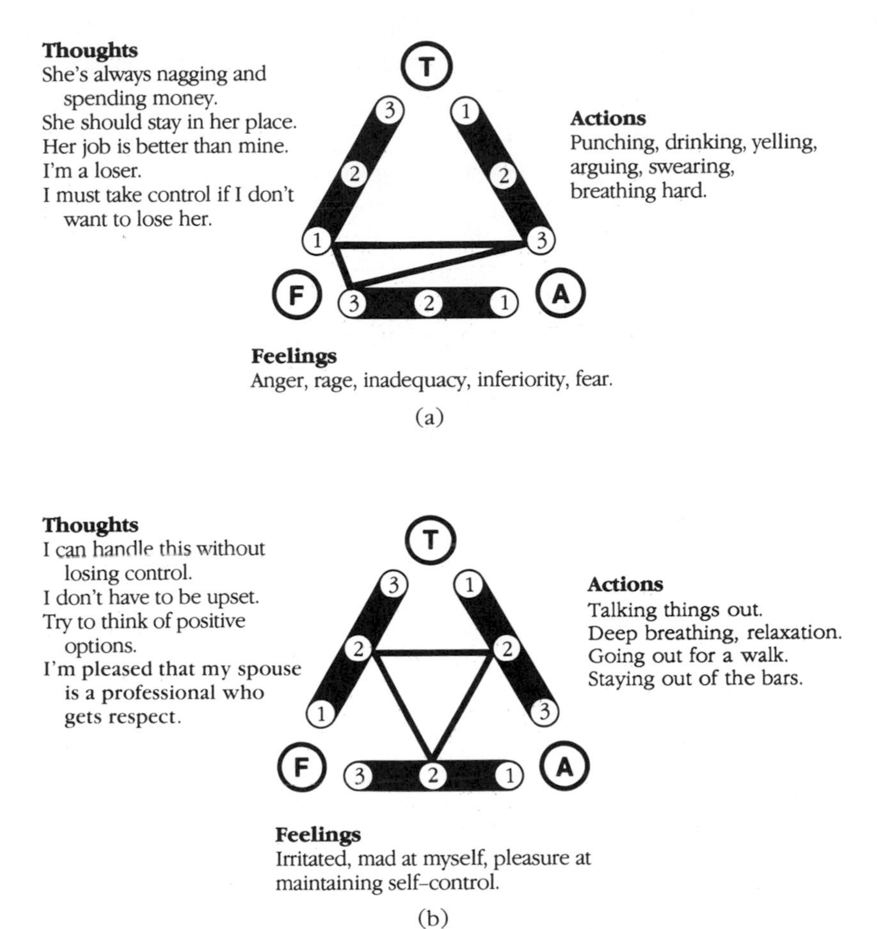

FIGURE 13.1 Client's behavior before and after participating in a helping relationship designed to stop spouse-abusive behavior.

Evaluation as a corrective process

Evaluation is corrective because it points out procedures that should be changed to help the client work toward realistic and concrete goals. In the exchange that follows, the client is helped to evaluate what happened, thus opening up alternatives that might be more helpful in the situation:

- HELPER: Last week you were going to write down the activities you did every half-hour to find out how you actually used your time. How did that go?
- CLIENT: Well. . . uh, not so well. You see, with the work that I do, it wasn't really possible to do this on three of the days I worked.

- HELPER: All right. Share with me what happened on those three days, and let's see if you can come up with something that will work for you on the job.

In this case, the evaluation process was *corrective* because it was discovered that the initial procedures were not working. Procedures were examined with the intent of designing them more effectively so the client would have a high probability of completing them (see Chapter 12). Evaluating procedures as the client tries them out can be very useful. Sometimes neither the client nor helper can anticipate specific events or problems that may interfere with the process. By looking at what is happening—or not happening—the helper can design procedures that have a greater chance of working in the client's specific situation. The helper constantly keeps in mind the change cycle as the helping process evolves by asking three questions:

Why is the client here?
What does the client want to do?
How are the goals to be achieved?

In other words, these essentially are

Why are we here?
Where are we going?
How are we going to get there?

The "we" questions suggest a mutual evaluation by both the client and helper.

Using the client's assessment as part of the evaluation

In a series of continuing interviews, periodically assessing how things are going *from the client's point of view* is essential. Although progress may be apparent from specific behavior changes being made, there may be other things occurring that may not be obvious. The guideline here is deceptively simple: If you want to know something, ask!

- HELPER: How do you feel about how things have been going in relation to your concerns?

This general question can be made more specific to changes—or not—in behavior: thoughts, feelings, and actions related to self and others.

- HELPER: Related to your concerns, what is happening in your daily life?

The general question can be made more specific to the client's job; plans; relationship with family, friends, employer, employees; and so on.

There may be changes in what the person thinks about self and interactions with others, or events may have made a positive or negative impact on the person and situation. Specific kinds of inquiry about the client's behavior as well as the helping process can be part of the regular, ongoing process of evaluation. In the effective helping relationship, the helper continually evaluates what occurs in terms of people, events, and processes related to the client and the helping process.

Obtaining information from others

With some clients and situations, getting information from others to assess progress is desirable or even essential. Depending on the setting in which the helper is operating, the client must give permission in advance. In certain court-referred cases, a series of items may demand information from other people. In such cases, permission is usually part of a detailed document that is signed before the helping process is started. Some strategies in Part 4 of this book require written contracts among the client, helper, and others. Parents, teachers, and probation or custodial people may be partners in working with the client to change behavior. Information from these people may be critical to evaluating individual's progress.

Evaluation as a summative process

A *summative evaluation,* occurring at the end of the helping process, makes overall, or summative, judgments about the worth or value of the helping process as it relates to the client's behavior. One way of approaching a summative evaluation is to review the process as a whole and ask questions such as these:

1. How effective was the process for the client? What changes did the client make?
2. What would I do differently and why? What was the effect of theories, techniques, and strategies used?
3. What have I learned as a result of this experience?
4. What are implications for working more effectively with others in the future?

Termination: A by-product of evaluation

Ward (1984, p. 22) summarizes primary functions of termination in the helping process:

1. Assessing client readiness for the end of counseling and consolidating learning
2. Resolving remaining affective issues and bringing about closure of the client–counselor relationship
3. Maximizing the transfer of learning and increasing the client's self-reliance and confidence to maintain change after counseling has ended

Others have written extensively about issues involved in termination: loss, separation, and grief (Lacoursiere, 1980); closure and sadness (Cohen & Smith, 1976); and grieving and leaving (Ward, 1982). However, most of these kinds of issues relate to much longer-term relationships that go beyond the intent of this book.

Useful signs that may suggest approaching termination can lie outside what the client says, including difficulty finding substantive things to discuss, forgetting to do homework, boredom in the interaction (Kanfer & Schefft, 1988), a reduction in the intensity of work, being late for appointments, joking, and intellectualizing (Corey et al., 1992). It may be that termination is nearing. We believe, however, that it is important for the helper to be aware of such cues *at any point in the interview process* and to deal with such behavior directly. This is a part of a continuing formative evaluation besides cues to termination.

Client readiness to leave the relationship

As termination approaches, several criteria are useful in evaluating client readiness to leave the helping relationship. Maholick and Turner (1979, pp. 588–589) suggest nine broad areas, which we have modified in the form of questions:

- Have initial problems or concerns been reduced or eliminated?
- Has client stress or pressure on the client been reduced?
- Can the client cope better with specific problems?
- Has the client's self-understanding improved?
- Does the client appreciate and respect differences in others involved in the situation?
- Can the client interact more effectively with other people, including giving and receiving love?
- Is the client being more responsible in everyday activities and interactions?
- Can the client obtain pleasure from everyday activities and positive interactions with others?
- Does the client feel able to live effectively without counseling?

Satisfactory answers to questions such as these provide strong indicators that the client may be ready to terminate the helping relationship.

Generally, the helping relationship is terminated for one of these three reasons:

1. *Referral is made* to a professional who has the skills and competencies to work more effectively with the client and situation.

2. *Goals of the helping process have been achieved.* One of the primary reasons the client sought help was because of specific concerns. The goal was to resolve the concerns by achieving personal goals. If the client has achieved goals, there is no reason to continue the relationship unless there are other

concerns the person wishes to pursue. Thus, the helping relationship is terminated when the helper and client mutually agree that goals have been achieved.

3. *There has been a change in situation or motivation, or no progress is being made.* Sometimes circumstances in the client's life bring about changes that alter things radically: a major illness, death, relocation, job, career, or family changes that result in the client leaving the helping relationship. If the client is not motivated to work on changing certain behavior, determining reasons for this reluctance is useful. Such a situation is depicted in the following example:

- HELPER: When you came in, you said you really wanted to learn how to manage your finances. In the last two weeks, you haven't kept track of expenses or recorded how you've spent money. I'm wondering if your situation or motivation to work on this area has changed since we started.

Here the helper has confronted the client about possible changes in the situation or motivation. The interaction that follows would then explore the client's situation and result in a mutual determination about what to do.

Reviewing the problem-solving process

In those cases where the helping process has been effective, reviewing the problem-solving process with the client is highly desirable. Its purpose is to help the client learn a general, self-correcting process, which can help more independent functioning in the future. The review is simply an overview of the helping process from the initial contact through resolving concerns and achieving goals. Four main areas of the relationship should be covered in the review process:

1. Identifying the problem situation for which the client sought assistance: What was the specific problem situation?
2. Assessing key aspects of the client's behavior in the situation: What were critical thoughts, feelings, and actions? Why and how were they demonstrated? What were important aspects of the situation (who was involved, where, and when)?
3. Selecting realistic, achievable goals to resolve the concern: What did the client want to happen?
4. Designing concrete, systematic procedures to achieve goals: How was behavior change accomplished?

Following this brief review, the helper asks the client to identify one or two other areas of life where this process might currently be applied. They briefly walk through the problem-solving behavior cycle to review the process. This walkthrough helps the client generalize the process and apply it to other specific situations.

Finally, when terminating the relationship, leaving the door open for future contacts, should they be needed, is desirable:

- HELPER: In the future, if you have concerns with which I can help you, feel welcome to come back.

This helps ease the client into another helping relationship if the need arises. Further, it signifies that there are times in everyone's life during which the helping process can be beneficial.

Follow-up

The *follow-up* process is initiated just prior to terminating regular interviews. It has two primary goals. The first goal relates to the client's behavior. Probably the single, most desirable reason for follow-up is to *reinforce positive behavior changes.* With any new behavior, some rough times occur, and follow-up sessions can help the client make it through these times more smoothly. Also, when problems do occur, the helper can provide *corrective feedback* so that the client can get back on track more readily than if doing it alone.

The second goal for follow-up relates to the helper. In the helping profession, we have a lot to learn about questions such as

- How permanent are the changes that occur in helping relationships?
- What kinds of techniques and strategies work most effectively in helping the client change behavior?
- Are some approaches to helping more effective than others? If so, which approaches, for what clients, with what problems, in which kinds of situations?

Follow-up can help provide data the helper can use to evaluate the effectiveness of the helping process.

Accountability and the helper

Throughout this book, we have taken the position that a helping relationship is much more than two persons getting together and talking. Rather, the helping relationship has a major goal of assisting the client in making specific behavior changes that will allow a more effective and satisfying life. We strongly believe that the helper has a responsibility to demonstrate the results of what is done in a professional helping relationship.

We have outlined several methods of keeping track of what happens through formative and summative evaluation, comparing client behavior when entering and terminating the relationship, and follow-up. Certainly, the clinical data on the kinds of changes made by the client are the key variables. Along with this, however, is the need to be more specific about what the helper does to assist in this process. How do theories, helping skills and strategies, and the interpersonal relationship affect the client? We have suggested some concrete ways of keeping track of what happens and how. As you develop, we strongly encourage you to assess your own work: What works, with which people, with what problem, when, and why?

Summary

1. Evaluation is the act of making judgments about the value of the helping process in light of the client's problem.
2. Evaluation implies assessing behavior against some criterion, or standard.
3. A formative evaluation is an ongoing assessment of progress in terms of the client's behavior and the helping process.
4. A summative evaluation is a review of the overall effectiveness of the helping process in helping the client change behavior.
5. Termination is a by-product of evaluation.
6. Termination occurs if the client is referred somewhere else, if client goals have been achieved, or if there have been changes in the client's situation or motivation that make it undesirable to continue the relationship.
7. Comparisons of the client's TFA triad and descriptive behavior on the TFA triangle (behavior blueprint) at the start and termination of the helping process is useful in assessing effectiveness.
8. During the final phase of termination, the problem-solving process is reviewed. The helper encourages the client to think about and apply the problem-solving process to other concerns.
9. When terminating a relationship, the helper leaves the door open for future contacts with the client should the need arise.
10. Follow-up can be useful in (a) reinforcing client gains or progress, (b) monitoring and modifying procedures to make them more effective, and (c) evaluating the effectiveness of the helping process as a whole.
11. The helper is responsible for documenting evidence about what works, with which persons, with which problem, when, and why.

Practice

1. Find a person who agrees to help you practice evaluating, terminating, and following-up.
2. Select role-playing situations from those that follow.
   a. Practice the evaluation, termination, and follow-up skills in five- to ten-minute mini-interviews.
   b. For each practice, use *relevant aspects* of the "Focus Checklist" in this chapter.
   c. When you have finished, write a brief summary in which you outline strengths, limitations, and suggestions for improving your performance in each area.
3. Try some mini-interviews (ten to twenty minutes in length) in which you try to appropriately integrate all helping skills and techniques

you have learned up to now. Use the Behavior Summary Checklist at the end of this chapter.

Role play

1. The client has been seen six times and has made substantial changes in behavior, achieving goals of assessing interests and aptitudes, exploring personal and career interests, as well as other options. The client's initial concern was career indecision, and the goal was to assess self and career options in a realistic manner. In your judgment, helping goals have been achieved.
   a. Practice summarizing the problem-solving process to date (initial concerns, goals, and procedures) and evaluate its effectiveness with the client.
   b. Practice terminating the interview, leaving the option open for additional contact in the future.
   c. Practice setting up a follow-up session to check on progress about six weeks after this interview.
2. You have had four sessions with this client who claims to be interested in improving relationships with others in the family but who has done very little to change interactions with them. (Assume that your client is *not* interested in continuing in the helping relationship.) You suspect there has been a major negative change in the client's motivation.
   a. *Confront* your client about the apparent lack of progress (see Chapter 9 for review if necessary).
   b. *Terminate* the helping relationship while leaving the door open for future contacts.
3. The client was rather shy in the initial part of the helping process but wanted to meet people more readily. Construct a TFA triad for the person (Figure 13.2) and *add* specific behavior descriptions at each T, F, and A vertex. Then, working with your coached client, review the progress made or *not* made and discuss implications for terminating or continuing the helping relationship.

Focus checklist

1. Is the client's problem resolved?
2. Are the client's goals achieved?
3. What are specific indicators of client gains?
   a. TFA triad shape is different.
   b. Behavior descriptions (blueprints) are different.
   c. Verbal feedback from the client.
   d. Observations of the helper about the client's behavior.
   e. Behavior assessments from others.
   f. Other observations: ______________________________
4. Has stress or pressure on the client been reduced?

Client's thoughts: ______________________________

____

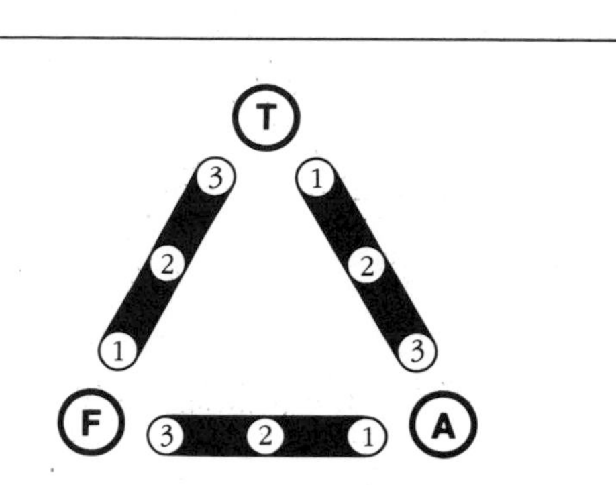

Client's feelings: ______________________________ ______________________________

Client's actions: ______________________________ ______________________________

FIGURE 13.2 A TFA triangle to be completed with client thoughts, feelings, and actions in a specified situation.

5. Can the client better cope with specific problems?
6. Has the client's self-understanding improved?
7. Does the client respect differences in others involved in the situation?
8. Does the client interact more effectively with others?
9. Is the client demonstrating more responsible actions?
10. Is the client obtaining pleasure from activities and interactions with others?
11. Is the client able to live effectively without the helping relationship?
12. What aspects of the client's motivation have changed?
13. What evidence suggests that the helping relationship is ineffective (from the client's behavior outside the interview setting to interpersonal behavior within the interview)?

Summary checklist

How many of the following elements did you use, and in what ways did you use them during your more integrated mini-interviews?

1. Structuring the helping relationship
2. Focus on thoughts, feelings, actions, or TFA combinations
3. Listening for understanding and nonverbal aspects
4. Asking open questions
5. Asking closed questions
6. Systematic inquiry
7. Clarification
8. Reflection

9. Use of silence
10. Confrontation
11. Client concerns
12. Goals
13. Procedures
14. Evaluation
15. Termination and follow-up

Instructions

To assess your interview skills, make a worksheet similar to the Behavior Summary Checklist. In the *left column* on the checklist, make a transcript of key words and phrases or sentences used by you and your client during the mini-interview. Then, *for each* transcript response, put a check in columns that are relevant. You might check from one to several items for each response.

When you have completed the Behavior Summary Checklist, review it carefully. Look for patterns where you emphasized or ignored content. Then write a brief summary of the mini-interview outlining your strengths, limitations, and suggestions for improving skills.

References and recommended reading

Bundy P. P. (1991). *Behavior assessments of pregnant adolescents using TFA systems.* Unpublished doctoral dissertation, Virginia Polytechnic Institute and State University, Blacksburg, VA.

Clow, D. R., Hutchins, D. E., & Vogler, D. E. (1990). Treatment for spouse abusive males. In S. M. Stith, M. B. Williams, & K. Rosen (Eds.), *Violence hits home*. New York: Springer.

Cohen, A. M., & Smith, R. D. (1976). *The critical incident in growth groups: Theory and technique*. La Jolla, CA: University Associates.

Corey, G., Corey, M. S., Callanan, P. J., & Russell, J. M. (1992). *Group techniques*. Pacific Grove, CA: Brooks/Cole.

Cormier, W. H., & Cormier, L. S. (1991). *Interviewing strategies for helpers* (3rd ed.). Pacific Grove, CA: Brooks/Cole.

Egan, G. (1990). *The skilled helper* (4th ed.). Pacific Grove, CA: Brooks/Cole.

Gunzburger, D. W., Henggeler, S. W., & Watson, S. M. (1985). Factors related to premature termination of counseling relationships. *Journal of College Student Personnel, 26*, 456-460.

Kanfer, F. H., & Schefft, B. K. (1988). *Guiding therapeutic change*. Champaign, IL: Research Press.

Lacoursiere, R. B. (1980). *The life of groups: Group developmental stage theory*. New York: Human Sciences Press.

Maholick, L. T., & Turner, D. W. (1979). Termination: That difficult farewell. *American Journal of Psychotherapy, 33*, 583–591.

Roffers, T., Cooper, B., & Sultanoff, S. M. (1988). Can counselor trainees apply their skills in actual client interviews? *Journal of Counseling and Development, 66*, 385–388.

Summary Behavior Checklist

Key words and phrases of the Client (CL) and the Helper (H)

Focus or Emphasis		
	T	
	F	
	A	
Structure		
Nonverbal aspect		
Open question		
Closed question		
Systematic inquiry		
Clarification		
Reflection		
Confrontation		
Use of silence		
Client's concern		
Goals		
Procedures		
Evaluation		
Termination		
Followup		
Strategies		

Tieman, A. R. (1991). *Group treatment for female incest survivors using TFA systems.* Unpublished doctoral dissertation, Virginia Polytechnic Institute and State University, Blacksburg, VA.

Ward, D. E. (1982). A model for the more effective use of theory in group work. *Journal for Specialists in Group Work*, *7*, 224–230.

———. (1984). Termination of individual counseling: Concepts and strategies. *Journal of Counseling and Development*, *63*, 21–25.

Worthen, B. R., & Sanders, J. R. (1987). *Educational evaluation: Alternative approaches and practical guidelines*. White Plains, NY: Longman.

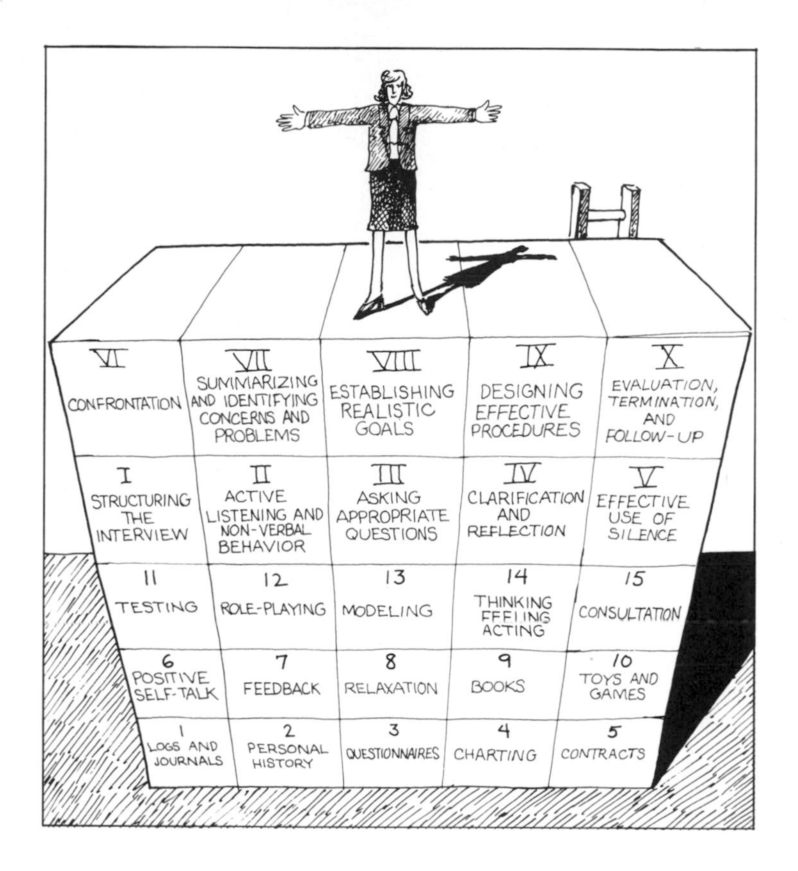

Strategies for Changing Behavior

PART FOUR includes a number of major strategies in the helping process. These strategies, combined into nine concise chapters, help the client move from current concerns to problem resolution.

FOUR

● Running Accounts and Recording Behaviors

14

Key points

1. Recording thoughts, feelings, and actions in a log or journal enables the client and helper to understand the dimensions of behavior, including frequency, duration, reactions, and other specific data.
2. A log or journal helps the client record thoughts, feelings, and actions at the time of an event without vagaries and distortions caused by memory lapses when client and helper meet at a later time.
3. Clients become aware of and monitor their own behavior by recording in logs or journals.
4. Guided fantasy used in combination with the journal approach can unlock feelings and suggest new actions or produce new thoughts about old solutions.
5. The act of writing a log or keeping a journal can be therapeutic as an activity by itself.

MANY people keep diaries as youngsters, and some continue the process as adolescents or adults. Helpers recognize the value of reading a client's thoughts, feelings, and actions described soon after they occur, without the almost inevitable distortions caused by lapse of time.

Recording thoughts, feelings, or actions on paper gives both the client and helper an opportunity for insight into how, when, why, where, and under what conditions particular behaviors occur. This strategy involves the systematic recording of the client's thoughts, feelings, and actions. For example, a client may be asked to commit thoughts and feelings to paper daily—to review the past twenty-four hours each evening before bedtime, perhaps—or to record personal reactions each time a particular event occurs, such as every time an ex-spouse calls to arrange visits with the children. Recording an emotional episode as soon as possible after it happens can give valuable detail to the helper at the next client-helper session.

- CLIENT (talking to helper): When my department head reprimanded me in front of the other designers, I was so hurt that I had to leave the room. I went to the restroom and cried.

If the client did not log this behavior at the time it happened, recall of a specific reaction might be difficult during a scheduled appointment. The client might remember the unhappiness but forget the intensity, the specific trigger (what was said), or the action that followed the unhappiness. Accurate data about the client's behavior provide a baseline from which to discuss what happens and why. Also, specific details can be learned about the situation, including *who* is involved, and *where* and *when* it occurs.

A client learning a new behavior might record reactions after each practice of that behavior.

- CLIENT: I forced myself to go up to an unknown family at church and introduce myself to the woman. My hands were sweaty, and I thought my voice wouldn't work, but I did it! And when she smiled, I suddenly realized from the look of relief on her face she was scared about not knowing anybody to talk to, either.

The log entry for this event might look like:

Thoughts: Thought I'd sound dumb and socially awkward; thought everyone else already knew what to say. Thought I was the only one who didn't know everybody else there.

Feelings: At first, fear of being silly, not knowing what to say. After I'd talked to her, I felt really glad I had because she seemed happier, too. It was easier than I thought it'd be.

Action: Introduced myself to Mrs. F. at church last Sunday.

It is often a good idea for the helper to provide some *format,* or *structure,* for the client. At the very least, the helper must be clear about *what* is to be recorded, *when* and *where*. Going through a typical situation and specifying what the client would do, how often, and where are good ways of practicing the behavior and ensuring that the client understands the process. Sometimes the helper must provide the client with a notebook and pen with which to record data. A technologically sophisticated client may choose to use a computer each day at home or at the office or a laptop computer for recording behavior "on the road." A pocket-sized electronic memo pad or calendar may also provide the means for recording behavior.

Use of the running account

A *running account* is a description of thoughts, feelings, and actions recorded soon after they occur and at fairly frequent intervals. This can be useful to both helper and client. Whether taken at given intervals (e.g., after meals, morning and night, just before a certain activity) or recorded as events occur, the running account has five main uses:

1. A running account gives the helper specific information about what happened and how the client reacted at the time of recording. Thus, intervening time and events that occur before the next meeting with the helper are less likely to influence the client's recollections.

- CLIENT: First, my boss asked me to get her coffee. Then she asked me to run out for a newspaper and to water her plants. Several times a day she asked me to do things like that for her, and I wrote them down so I would remember exactly what happened. I was furious, especially by the third day, but I haven't said anything to her yet.

2. A running account causes the client to focus on particular behaviors and asks for accurate recording of these behaviors. The client's observations are likely to be more reliable if recorded at the time something happens. If several days elapse between the event and discussion of the event with the helper, observations may be vague.

- HELPER: Right after you and your roommate have had a fight, I want you to record what happened: What she said and how she said it. I also want you to write down your *thoughts,* your *feelings* at the time, and what you *did* or *did not do*.

3. A log or journal offers clients the means for monitoring their own behavior. It provides daily contact with the problem-solving process by catching specific thoughts, feelings, and actions related to the problem while they are fresh in the client's mind. Asking the client to dredge up information from memory during a once-weekly meeting may produce inaccuracies.

- CLIENT: When I went back over my log, I realized I had slept better when I had jogged in the morning and done relaxation exercises during the day.

4. Recording thoughts, feelings, and actions can be therapeutic simply by expressing them on paper or screen. The cathartic effect of flushing out thoughts and reactions is reason in itself for daily recording, even if no further use is made of the material.

- CLIENT: My anger seemed silly—even funny—when I put it down on paper. I seemed to be looking for things to be mad about, even making some of them up.

5. The act of writing down specific behavior and events can provide the client with awareness about otherwise unnoticed aspects of the situation. Accurate awareness is the first step in the process of change.

- CLIENT: I never knew I had so many interruptions at work until I logged them. No wonder I feel so frustrated at finishing so few things—I'm always letting myself be interrupted. I'm not so much a slow worker as an interrupted one. Maybe I'm better than I thought at what I trained to do.

Ties to the change cycle

Using a running account will be most useful for the *assessment* of the client's current status, learning how the client thinks, feels, and acts. Especially when the helper works with the client to determine concrete aspects of behavior, charting specific dimensions of positive and negative thoughts, how feelings are affected, and associated actions of the client can be used to assess triggers, or stimuli, that result in certain behavior. *Goals* can be specified in terms of changes from one's current behavior to more desirable behavior. Running accounts can help detail increased or decreased specific behavior over a period of time. Related to this, *procedures* for changing behavior can be monitored so that both the client and helper can determine whether certain techniques and strategies are having the desired impact. The client's behavior can be monitored on a daily or weekly basis, and the client's log can alert one to changes that may need to be made.

These strategies are useful to almost any client. However, they are especially important when the client is *not* cognitively aware of what happens, why, and under what circumstances.

Guided fantasy

A specialized use of the journal, well established in counseling literature, can be made through *guided fantasy*[1] (Progoff, 1975), in which a client may, for example, construct a scenario on paper and test out behaviors before actually trying them in action. An event can be rewritten to include actions and consequences that *might* have taken place, as well as those the client actually did use. By writing personal history in a direction hoped for in the future, the client can, in effect, create a new self-image.

The professional helper may also verbally lead a client through a guided fantasy, in which the client writes down reactions, as in a response to unresolved events of the past. Reliving past events and fantasizing different outcomes may clarify decisions to be made in the future, relieve guilt, or confound some long-held rationalizations. For example, through guided fantasy, a client who blames her husband for limiting her options in the past may find that she can, in fact, make decisions about her own future; that she, not her spouse, has made her choices in the past; and that her private fantasies, taken to a logical conclusion, might not be as rosy as she thought:

- CLIENT: When I followed through in my mind what my life might have been, I guess I made some good decisions in the past. What if I'd married John instead? Look at him now: kids in trouble, an alcoholic, bouncing around among wives and jobs.

Limitations of the log or journal

The log or journal method of recording behavior is similar to other written strategies:

1. The client must be comfortable with written language for it to be useful.
2. The client must be diligent and accurate in recording information at various times. If the client writes the entire account just before the helping session, the purpose of the exercise is defeated. In such a case, it would be good to explore reasons for this. It may be part of other unresolved aspects that must be examined, including the client's motivation.
3. The client has a chance to edit content presented to the helper.
4. Nonverbal indicators cannot be discerned from the printed page. However, a written description of behavior is a chance for some

[1]Do *not* use this technique without further training. It is described here only to give beginning helpers an idea of a strategy they may want to study in depth.

people to express more private thoughts and feelings than they can articulate in a discussion.

5. Some people have difficulty with fantasy; using the journal approach for guided fantasy will not be useful for clients who cannot or will not fantasize effectively.

Case study: Using the journal

As a young engineer, Lara Fuentes loved her work, felt supported by most of her colleagues in the plant, and really wanted to retain her position. However, she felt she had to improve communication with her supervisor or change to another job. "If something can't be changed, I want a transfer or I'm quitting! I'm not going back to that engineering department the way things are!" she exploded in the personnel office.

Mark Davis, the personnel manager, heard her out (sometimes the best method of dealing with a client's anger is to just let them talk or ventilate—to get it out of their system). He then suggested that she keep a log of incidents that disturbed her and her reactions to them during the week, using the pocket electronic memo-keeper she always carried. Ms. Fuentes agreed to continue to work in the department and to meet with Mr. Davis the following week, bringing her log for discussion.

At their second meeting, Ms. Fuentes produced a description of nine incidents that had upset her within the past week. After an analysis, she and Mr. Davis concluded that four of the incidents related to times the supervisor seemed to feel challenged or threatened. All involved new ideas or work in areas in which Ms. Fuentes's expertise was clearly superior to her supervisor's. Two incidents seemed ridiculously trivial to Lara when she reread her accounts, and she dismissed them as overreaction on her part. Two other incidents seemed to involve someone else in the department, and Ms. Fuentes could see that she was really the recipient of deflected anger toward someone else. One incident seemed an important complaint.

By analyzing the events and her responses, Ms. Fuentes realized that she could predict some reactions, avoid some, and discount others. This gave her information she thought would enable her to *plan her interaction* with her supervisor in a way that would give her more control over the communication. Ms. Fuentes agreed to continue her work now that her problem had been defined more clearly. And Mr. Davis, armed with facts, determined to try to help the supervisor improve managerial skills.

Analysis

Reading Lara's accounts in a journal triggered the following:

1. The personnel manager (the helper in this setting) could estimate the frequency and severity of the problem. Data recorded in the journal

established just what the incidents were, how often they happened, and who was involved in each.
2. The client could monitor her own reactions and gain perspective from a less emotional viewpoint. She could observe that at least twice she had overreacted.
3. Both helper and client gained facts about the situation so that problem solving could begin. Until there is a common understanding of the problem, planning strategies is difficult.
4. The client could express her feelings freely on paper, thus relieving some of the intensity of a confrontation with her supervisor.
5. The client was willing to let some time elapse without a hasty solution being tried, yet she felt that something was being done—resulting in a cooling-off period to allow for data gathering.

Practice

1. Find a person who agrees to help you learn how to use the strategy of asking a client to record behavior in a log or journal. Assume that you and your coached client have previously identified the client's concerns.

2. Select one situation from the following role plays. Before starting your mini-interviews, clarify specifics of the situation: age and gender of the client, setting of the interview, and other aspects essential to getting started. Assume that your client has sufficient writing skills to keep a reasonably detailed log or journal.

3. Conduct a ten-minute mini-interview for each of the role-playing situations.[2] Record your interaction or have another person observe the interview, especially noting aspects related to the description of instructions for writing and to discussion of the client's written work. Start the interview by summarizing the client's concern (done by either the client or you).

4. In role-playing situation 1, discuss with your coached client how to record thoughts, feelings, and actions in a log or journal. In role-playing situation 2, use the client's log or journal to learn about thoughts, feelings, and actions relative to the role play. Discuss with the client those thoughts, feelings, and actions and how you see them relating to the concern you have discussed together.

5. Following each mini-interview, ask your coached client for *feedback* about the interaction. Then listen to the tape and answer the questions in the "Focus Checklist," which follows the role-playing situations.

6. Try some mini-interviews in which you practice appropriately integrating the techniques you learned in Parts 1 and 2 with use of the log or journal. Create your own role-playing situations relevant to your setting or use any of the situations presented in the first two parts of this book. Use the "Summary Checklist" and record all appropriate data from these more inte-

[2]Where the role-playing situation calls for use of a log or journal, have your coached client write out whatever is necessary before beginning the mini-interview.

grated interviews. The "Summary Checklist" and its instructions follow the "Focus Checklist."

Role play

1. You are working with a couple who want to reduce conflict. You need to learn just what kinds of conflicts are occurring, at what times, on what topics, and under what conditions. You ask both husband and wife to write separate accounts of each angry encounter that occurs during the next week. Explain to them what to record and discuss how to record it and what you will do with the information next week.

2. Continue with the preceding situation, discussing information one of your clients has brought to you. Using a log your coached client has prepared, ask the client to draw conclusions that lead to some insights into the couple's conflict.

Focus checklist

1. From what the client had written, what useful information did you gain about the client's thoughts, feelings, and actions?
2. Did you give clear instructions to the client about how to keep a log or what to write in a journal?
3. How did you relate what was in the log or journal to the client's present concern?
4. Did you and the client discuss a method for logging, or entering, into a journal? Was it clear to the client what, when, and how to record?
5. Did you discuss how irrelevant, inappropriate, or confusing aspects in the log or journal are discarded, revised, or clarified?

Summary checklist

How many of the following elements did you use, and in what ways did you use them during your more integrated mini-interviews?

1. Structuring the helping relationship
2. Focusing on thoughts, feelings, actions, or TFA combinations
3. Listening for understanding and nonverbal aspects
4. Asking open questions
5. Asking closed questions
6. Systematic inquiry
7. Clarification
8. Reflection
9. Use of silence
10. Confrontation
11. Client concerns
12. Goals

13. Procedures
14. Evaluation
15. Termination and follow-up

Instructions

To assess your interview skills, make a worksheet similar to the Behavior Summary Checklist. In the *left column* on the checklist, make a transcript of key words and phrases or sentences used by you and your client during the mini-interview. Then, *for each* transcript response, put a check in columns that are relevant. You might check from one to several items for each response.

When you have completed the Behavior Summary Checklist, review it carefully. Look for patterns where you emphasized or ignored content. Then write a brief summary of the mini-interview outlining your strengths, limitations, and suggestions for improving skills.

References and recommended reading

Fox, R. (1982). The personal log: Enriching clinical practice. *Clinical Social Work Journal, 10*(2), 94–102. This article provides an extensive description of the clinical use of the client's log, including guidelines for using the log.

Gilliland, B. E., James, R. K., Roberts, G. T., & Bowman, J. T. (1984). *Theories and strategies in counseling and psychotherapy*. Englewood Cliffs, NJ: Prentice-Hall. The discussion of use of fantasy in counseling in Chapter 5 includes gestalt therapy and especially fantasy that leads to emotional resolution of an unfinished situation.

Paisley, R., Gerler, E. R., Jr., & Sprinthall, N. A. (1990). The dilemma in drug abuse prevention. *School Counselor, 38*(3), 113–122. The authors used journals with adolescents to explore issues of substance abuse.

Progoff, I. (1975). *At a journal workshop*. New York: Dialogue House. This book offers a complete discussion of the use of the intensive journal and guided fantasy as counseling tools.

Summary Behavior Checklist

Key words and phrases of the Client (CL) and the Helper (H)

T F A

Focus or Emphasis: T	
Focus or Emphasis: F	
Focus or Emphasis: A	
Structure	
Nonverbal aspect	
Open question	
Closed question	
Systematic inquiry	
Clarification	
Reflection	
Confrontation	
Use of silence	
Client's concern	
Goals	
Procedures	
Evaluation	
Termination	
Followup	
Strategies	

Autobiography and Personal History

15

Key points

1. An autobiography is a chronicle of a client's life; a personal history covers a portion of life and may include more information than does the autobiography.
2. A client can record an autobiography or some portion of life history to give the helper insight into events such as personal events (marriage, divorce, death) or external events (death in the community, war) and into the client's thoughts, feelings, and actions surrounding those events.
3. The client may describe a particular aspect of life such as progress of a career or marriage, parental experience, or some other single experience.
4. Writing or recording an autobiography can be cathartic, freeing the client's mind of some emotional baggage.
5. By asking the client to write on paper or record on audiotape or videotape a conceptualization of the concern as well as goals that should alleviate that concern, the helper can determine whether helper and client are communicating accurately.
6. Referring to a previously recorded personal history enables the helper and client to recognize and measure changes in the client's thoughts, feelings, and actions over time.

EXPRESSING one's thoughts, either verbally or written, often gives substance and form to shadows in the mind. Vague ideas become more concrete. An *autobiography* is a listing of life events, often from earliest memory to the present. A *personal history* is a segment of autobiography, which may focus on a single short time period (e.g., the illness and death of a spouse) or on a larger portion of life (e.g., employment history). The writer can simply detail events or can give a fuller account by including feelings about and reactions to events: events plus interpretation and commentary, actions plus thoughts and feelings.

Uses of the autobiography or personal history

Recording an autobiography or a personal history can be useful to a client in several ways:

1. It clarifies events, feelings, and trends for the client, allowing perspective on life or a portion of it by seeing events in progression on paper or hearing them on tape. For example, the client may list significant events in chronological order, in order of importance, in a stream-of-consciousness fashion as they surface in the mind, or in whatever other order is decided. Recording events or reactions can give the helper valuable insights into how the client views reality, as well as information about past history. A written or

verbally recorded statement may or may not be accurate, but it surely reveals the client's view of an event:

- CLIENT: I was fired because my boss was totally unreasonable.

This statement can then be explored to assess the reality and facts in the situation.

2. Often the act of recording words frees the client's mind of emotional baggage; the act of writing or recording can be cathartic, and the reality of recorded words can sometimes bring a matter of factness to feelings that were previously overwhelming. The written page or the tape also gives the helper and client a common base for discussion. Some people, *especially more introverted ones,* may express themselves more freely in writing than verbally to another person. Or they may prefer to speak in private to an audiorecorder or videorecorder. Some clients may find the impersonality of the computer screen, for example, a less threatening medium of expression than talking to the helper:

- CLIENT: When I was writing this last night, I almost couldn't hit the keyboard fast enough. The words just sort of jumped from my mind onto the screen.

A brainstorming, or stream-of-consciousness style, encourages expression without regard for spelling, grammar, organization, accuracy, or completeness. A word processor or a tape recorder is a convenient way to keep a contemporaneous account of events and thoughts as life events unfold, producing a kind of ongoing, unfolding autobiography.

3. A good way to check whether the helper and client are conceptualizing the same concerns and moving toward the same goals is for the client to put perceptions of the concerns and goals into writing. For some, expression comes easier on paper or on a keyboard than verbally, and for these clients in particular, communicating in written as well as spoken words is very valuable. The client can write, reflect, and ensure that the wording says as clearly as possible what the client wants to convey. Similarly, the helper can read and reflect on what the client has written about problems and pathways to solutions:

- HELPER: Your goal statement at the end of your autobiography is very clear—to move into a salaried job and to supervise.

Some clients may prefer to record their thoughts, feelings, and actions on audiotape or videotape. Whereas a videotaping system is somewhat cumbersome and relatively expensive, a small tape recorder can accompany the client almost everywhere.

4. Keeping the recorded autobiography, personal history, problem statement, or written goals of a client on file can be a good way for the helper to determine progress. This also lets the client see how thoughts, feelings, and actions change over time. After making progress through a helping rela-

tionship, the client may be surprised to see what was contained in the original statement of the problem:

- CLIENT: I thought I came in to talk about changing majors, but now I see that I need to think about my feelings about authority figures, especially my parents. Changing majors is all tied in with what I want to do and what I think they want me to do. I didn't begin to realize that until I wrote down what I felt about my schooling and then did some thinking about it.

5. Particularly when the client is maturing or when life changes are causing strong emotion, a client may be very surprised to see how perceptions change over time. Moderating emotion following grief, divorce, loss of job, or depression will cause gradual changes in a client's perceptions of life status that will perhaps not be apparent as time and the helping process go on. A teenager will be surprised (and often embarrassed!) to see how views change with age. (Remember how parents sometimes get smarter as children get older?)

- CLIENT: You mean I recorded that last year? That's not how I feel now! My feelings have really changed!

Limitations of the autobiography or personal history

The written or recorded personal history or autobiography has some significant limitations:

1. It is useful primarily with literate individuals who can express themselves adequately on paper or screen. Although some benefit could be gained by having a client dictate to another person who then writes, much of the effect of organization, reflection, and introspection that occurs when a literate person writes about self is lost. An articulate client not comfortable with writing may prefer to use audiorecording or videorecording to record thoughts, feelings, and actions.
2. The helper receives information only from the client's point of view. The client can edit the version given to the helper, eliminating the spontaneity more likely to be present in one-to-one communication.
3. Written narrative is unidimensional; the helper cannot read body language from a printed page or an audiotape, and emotional inflection is even more limited in a written account. Videotaped reflection would not have this limitation.
4. Chronology or actual data may be incorrect because of faulty memory or the rearrangement (deliberate or unconscious) of events in a client's mind.

Autobiography, personal history, and the change cycle

Writing an autobiography or personal history can be used in any of the three stages of the change cycle. In the first stage, where the objective is to understand the client's behavior and situation, an autobiography or personal history is useful in providing the helper with a "fast start." Much ground can be covered in a short time, by reading what the client has written. The act of writing enables the client to reflect on events and put these into perspective—a clarifying experience by itself.

Stage two (goals) of the change cycle is another area where the personal history and autobiography can be useful. Both client and helper can see *patterns* and discuss implications for goals, especially if the client is instructed to prepare the written material with goals in mind.

The third stage (procedures) of the change cycle may also use these written materials to help project steps the client may wish to take or explore. Such explanation can assist the client in projecting scenarios into the future—to mentally rehearse events that could occur fairly immediately or in the long run.

This strategy, as well as other written and recorded strategies, is extremely useful for both client and helper to learn about the past; to extract thoughts, feelings, and actions from the past; and to move into the future with resolution of some past problems. Such strategies can also be useful in planning for events and rehearsing possible strategies for the future.

Case study: Using the autobiography

Referred to a helper by his physician, Mr. Sampson was still grieving about his wife's death two years previously. In the initial interview with the helper, Mr. Sampson spoke of his wife as a saint and related how closely he had attended her during the six months prior to her death. To get an idea of the chronology of events and of Mr. Sampson's role in his wife's illness and care, the helper asked the client to write a history of Mrs. Sampson's last year and of his feelings about her illness.

When Mr. Sampson returned the following week, he showed signs of agitation. As he handed his narrative to the helper, he said,

> At first, I started writing about how my wife felt and what I had done when she got sick. Then I began remembering how angry I felt that this thing was happening to us and how I worried about what my life would be like when I was alone. I remembered how guilty I felt every time I knew she wanted me to stay with her and how glad I was for an excuse to go to the bank or the pharmacy or the gas station, just so I could be free for a short while. The freer I felt, the more guilt I felt. Then I'd tell myself what a terrible person I was because she couldn't do any of these things and I shouldn't be enjoying them either, much less being glad to be away from

her when we had such a short time left together. The hospice volunteers tried to talk to me about these feelings, but I denied having them then. When I began writing what you asked me to, those feelings just came flooding back, and it was almost as if my pen were writing without my help; the words just flowed out onto the paper. You may think I'm a terrible person to feel like that, but I really feel better after writing it all down.

The helper used Mr. Sampson's description of his emotions to probe further, assisting Mr. Sampson to see how his negative thoughts about himself were impairing healthier feelings and actions.

Analysis

The helper used Mr. Sampson's description to

1. Learn about facts in the past that were influencing present thoughts, feelings, and actions.
2. Understand how the client reacted emotionally to certain events.
3. Help the client clarify his present feelings about his wife's illness and death.
4. Decide where to probe further for feelings.
5. Ascertain that both the client and helper held the same view of past circumstances, for clear communication.
6. Give the client the chance to express his feelings freely and forcefully and rid himself of some of the previously unexpressed emotion that was troubling him.

Practice

1. Find a person who agrees to help you learn how to use the strategy of asking a client to write an autobiography or personal history. Assume that you and your coached client have previously identified the client's concerns.
2. Select one situation from the list of role-plays that follows these instructions. Before starting your mini-interviews, clarify specifics of the situation: age and gender of the client, setting of the interview, and other aspects essential to getting started. Ask your client to prepare a short autobiography on the topic of the situation you have selected.
3. Conduct a five-minute mini-interview for each of the role-playing situations. Record your interaction or have another person observe the interview, especially noting aspects related to using the client's autobiography as you suggested in the previous meeting.
4. Use the client's autobiography to learn about thoughts, feelings, and actions relative to the role-playing situation. Discuss with the client

those thoughts, feelings, and actions and how you see them relating to the presented concern.

5. Following each mini-interview, ask your coached client for feedback about the interaction. Then listen to the tape or talk to your observer and answer the questions in the "Summary Checklist," which follows the list of role-playing situations.
6. Try some mini-interviews in which you practice appropriately integrating the techniques you learned in Parts 1–3 with use of the autobiography or personal history. You can make up your own role-playing situations or use any of the situations presented in the first two parts of this book. Use the "Summary Checklist" and record all appropriate data from these more integrated interviews. The "Summary Checklist" and its instructions follow the "Focus Checklist."

Role play

1. Your client is a forty-year-old woman who has not previously held a job outside the home for pay and who, for the first time, needs to earn money to live. Your client is unsure of job skills and of personal characteristics necessary to hold a job. During the last interview, you asked the client to write a work autobiography, listing kinds of things done at home and in a volunteer capacity, along with feelings about performing those tasks and about working with other people both at home and outside.

2. A parent has brought in a thirteen-year-old son who has always had trouble making friends. The goal is to learn how to get along with peers. At your last session, you asked your young client to use a tape recorder to document what has happened in the past with attempts to establish friendships. He was to especially focus on the beginning of each school year and what it has felt like to enter a new school and neighborhood on the several occasions when the family has moved.

Focus checklist

1. What useful information did you gain about thoughts, feelings, and actions from what the client had written or recorded?
2. Did you ask how the client felt during and after writing or recording autobiographical thoughts?
3. Did you relate the *past* as described in the autobiography to the *present* as being experienced by the client?
4. What implications do the written data have for the future as described in the client's goals for the helping relationship?

Summary checklist

How many of the following elements did you use, and in what ways did you use them during your more integrated interviews?

1. Structuring the helping relationship
2. Focus on thoughts, feelings, actions, or TFA combinations
3. Listening for understanding and nonverbal aspects
4. Asking open questions
5. Asking closed questions
6. Systematic inquiry
7. Clarification
8. Reflection
9. Use of silence
10. Confrontation
11. Client concerns
12. Goals
13. Procedures
14. Evaluation
15. Termination and follow-up

Instructions

To assess your interview skills, make a worksheet similar to the Behavior Summary Checklist. In the *left column* on the checklist, make a transcript of key words and phrases or sentences used by you and your client during the mini-interview. Then, *for each* transcript response, put a check in columns that are relevant. You might check from one to several items for each response.

When you have completed the Behavior Summary Checklist, review it carefully. Look for patterns where you emphasized or ignored content. Then write a brief summary of the mini-interview outlining your strengths, limitations, and suggestions for improving skills.

References and recommended reading

Aust, P. H. (1981). Using the life story book in treatment of children in placement. *Child Welfare, 60*(8), 535–536, 553–560. The author discusses the therapeutic use of biography/autobiography for foster children.

Bergmann, S., & Creighton, T. (1987). Parent–student communication: A middle level school challenge. *Middle School Journal, 19*(1), 18–20. Parents and children write to each other to express things each wants the other to know.

Beste, H. M., & Richardson, R. G. (1981). Developing a life story book program for foster children. *Child Welfare, 60*(8), 529–534. A life story book—prepared by social workers, foster parents, and perhaps parents, as well as the child—can bridge gaps in foster children's lives.

Fox, R. (1983). The past is always present: Creative methods for capturing the life story. *Clinical Social Work Journal, 11*(4), 368–378. Fox presents several ways of eliciting autobiographical information from clients.

Sabatini, L. (1989). Preparing eighth graders for the social pressures of high school.

Summary Behavior Checklist

Key words and phrases of the Client (CL) and the Helper (H)

T F A

	Focus or Emphasis
	T
	F
	A
	Structure
	Nonverbal aspect
	Open question
	Closed question
	Systematic inquiry
	Clarification
	Reflection
	Confrontation
	Use of silence
	Client's concern
	Goals
	Procedures
	Evaluation
	Termination
	Followup
	Strategies

The School Counselor, 36(3), 203–207. Eighth graders develop brief autobiographies as part of orientation to high school.

Weinz, K., & McWhirter, J. J. (1990). Enhancing the group experience: Creative writing exercises. *The Journal for Specialists in Group Work, 15*(1), 37-42. The authors describe the use of creative writing in a group setting as adjunctive therapy in group work.

Structured Information Gathering: Questionnaires, Open-Ended Sentences, and Structured Interviews

16

Key points

1. Open-ended sentences (beginnings of statements that the client completes) can give the helper much insight into a client's thoughts, feelings, and actions.
2. Questionnaires elicit information from clients in a structured format.
3. Questionnaires and open-ended sentences may limit responses because certain topics or questions the client might like or need to answer are not included on the form.
4. Structured interviews give the helper information according to a predetermined format, with the chance for probes and follow-up questions to elicit more data on the client's thoughts, feelings, and actions.
5. Information garnered from structured writing and interviews can be *one* basis for combining clients with similar needs into helping groups.

A useful writing technique akin to autobiography and personal history writing, but much more structured, is the questionnaire or some variation thereof, such as open-ended sentences. As with the autobiography, the helper can learn much about thoughts, feelings, and actions from what the client writes on paper. The structured interview follows a similar format, but the helper–interviewer asks questions and records answers, instead of the client completing a written questionnaire.

Uses of structured writing

1. Questionnaires ask specific questions and are typically used to establish chronology of events, importance of activities, frequency of occurrence, or other such specific data. For example,

- What were five important things that happened to you [between the ages of 10 and 15; while employed with XYZ Company; in the experience you just described]?
- What are the three most important things [in your life; in looking for a partner; in leisure activities; in a job; in your family]?

2. Like the autobiography, questionnaires do not replace interaction with the helper but serve instead as a beginning point for discussion and as a means for allowing the client to gather data and to organize and explore *past* and *current history* of events and personal reactions to these events. Questionnaires can target either thoughts, feelings, actions, or various combinations of these. The structured-writing strategy can also be useful in drawing on past experiences as well as having the client *project* desired situations. For example,

- How did you *feel* about the relationship you had with your parents when you were growing up?
- What kind of *activities* do you like to do outside the work setting?
- What would you see for yourself in your career if you could predict the *future?*
- What will your life be like five years after you finish your education?

At times the helper may find it useful to orally ask open-ended sentences, rather than asking for written completions, turning what would usually be a structured-writing strategy into a structured interview.

3. A variation of the questionnaire especially useful in helping both children and adults articulate their feelings is a set of open-ended statements the client completes. Statements may include such items as these:

- I get really angry whenever my teacher . . .
- When I am with kids my own age, I . . .
- I am happiest when . . .
- I get nervous when . . .
- In my leisure time, I would like to . . .

Completion of such statements gives the helper entry into the client's thoughts and feelings. For the overextended helper in an institutional or agency setting, this sort of technique can be particularly useful as a *preliminary screening device* to decide who needs help with which kinds of concerns. Some agencies use such questionnaires or open-ended questions as an intake procedure for all clients. But again, a caution: Collecting information about thoughts, feelings, and actions is only a *part* of the helping process and does not replace the helping relationship that develops through interaction between helper and client.

- HELPER: Thank you for completing this questionnaire on study habits, Junie. Your answers tell me a lot about what you already know and what you believe you need to learn about study skills.

4. Responses to open-ended sentences or questionnaires can be indicators of which clients have similar needs and thus who can be included with whom for group work.

- HELPER: Jim, Bob, and Joel, I asked all three of you to come to our meeting today because each of you wrote on your questionnaire that you wanted to learn to stand up for yourself and not get upset when the other guys tease you. I think we can work on that together as a group.

As with any other *screening technique,* however, the helper must be aware that the client may not be fully revealing or accurate with statements made on paper. Especially with younger children, as well as some adolescents and adults, written skills may not be sufficient for complete expression of feelings.

5. Many published checklists and questionnaires are available to help-

ers. Many helpers use these assessment tools or similar ones they devise themselves as an initial intake instrument for new clients.

Uses of structured interviews

1. The structured interview allows the helper to collect predetermined data from the client. By asking the same questions of all clients, such as in an intake interview in an agency, the helper is sure to remember to collect all pertinent data about a problem, rather than risking forgetting something that may prove valuable later.

- HELPER: I'm going to ask you some questions we will be using to screen all potential applicants, so that we can be sure that we allow you to give us the same information as does every other applicant for this job.

2. Guidelines for drawing up a structured interview will be determined by the *intended use* for the information gathered. If the interview is an agencywide data-gathering instrument, it should be prepared by all those helpers who will be using the data once they have been gathered.

- CLIENT: If I understand you correctly, you will be asking me all the information the team needs, so that I won't be answering the same question each time I see another cardiac team member. Is that correct?

3. Predetermined questions mean that all helpers in a given setting will collect the same information on each client. As interviewers are trained to use the instrument, similar kinds of data will be solicited from each client.

- HELPER: I will be asking you the questions all of us [school social workers] ask parents of children who are being evaluated for placement in special education classes.

Limitations of questionnaires and open-ended sentences

Questionnaires and open-ended sentences, along with other written techniques, have clear limits to their usefulness.

1. To use the technique, clients must be comfortable with and capable of using written language. The client's social, educational, and cultural background must be considered in evaluating written responses to structured formats. Language itself may be a barrier; to gain complete responses, the questionnaire or sentence-completion form should be in the client's primary language.

2. Although responses are open-ended, structured questions or state-

ments may not cover all areas appropriate for a given client. Unless the helper knows how to combine this strategy with others to elicit information, valuable information may be overlooked.

3. As with other written techniques, the client has a chance to reflect on what has been written and to edit responses, precluding spontaneity and perhaps affecting truthfulness. Especially if the material for completion is presented on a computer screen, the client has considerable opportunity to edit responses.

4. Helpers are limited in assessing nonverbal client signals from the printed page, except as revealed by handwriting or client-response patterns to the questions or statements. When discussing written responses, the *helper can be especially aware of the nonverbal client behavior* that can provide additional insight into feelings.

Structured data gathering and the change cycle

In stage one (assessment), the helper can benefit from using structured data-gathering devices to help in understanding more about the client's behavior and situation. This is especially true when there is limited time and where the client may be relatively nonverbal. The technique is also particularly useful when the same information is needed on all clients, such as in a job-placement agency where certain kinds of data, such as education and previous work experience, are always needed. Knowledge gained about the client's thoughts, feelings, and actions in a specified situation can be followed up in later sessions. The helper can assist the client in seeing patterns and relationships between events or concepts. For example, the client may have revealed in answers to open-ended questions the kinds of situations that lead to anger or depression. In a structured interview, the client may have discussed patterns of achievement or leadership that emerged in childhood and continued into adult life. The client can be confronted when discrepancies exist between data and observations made in the interview. In addition, structured data gathering can become one way of evaluating changes that take place from the onset to the completion of the helping relationship.

Case study: Using the questionnaire

Mr. Sanchez was worried because Michael seemed so hostile and aggressive toward the other fourth graders in his class. He seemed to have the attributes necessary to make friends: He was bright, articulate, possessed of a good sense of humor, and well coordinated physically. The teacher asked the school counselor to try to help Michael learn how to use his wit and physical skill in some way other than sarcasm and fighting with other students. The counselor asked Michael to complete a set of open-ended sentences before their first interview. The counselor was not surprised to see Michael's responses (underlined) to many of the questions:

- When I am with other kids my own age, I am afraid they won't like me.
- I feel important when . . . never.
- I wish I had more friends.
- I am good at nothing.

Other answers similar to these confirmed the counselor's conjecture that Michael had a very poor view of himself and that, although Mr. Sanchez saw Michael's many strengths and assets, Michael himself was unaware of what these strengths were and how his strong points could help him make friends. The counselor saw a need to help Michael redefine his view of himself and to learn to use the many skills he had in different, more positive ways.

Analysis

By using open-ended sentences the counselor was able to

1. Confirm his impression that the client indeed had a low self-image.
2. Estimate the degree of negativism the client felt about himself: "I am good at nothing."
3. Have the client's own written perception of his peer relationship: "When I am with other kids my own age, I am afraid they won't like me."
4. Elicit a beginning goal statement for counseling: "I wish I had more friends."

The same kinds of assessments, observations, and procedures can be used by the helper just as effectively to gain information from adults who have poor self-concepts or other concerns that they bring to the helper.

Practice

1. Find a person who agrees to help you learn how to use questionnaires or open-ended sentences. Assume that you and your coached client have previously identified the client's concerns.
2. Select one situation from the list of role-plays, which follows these instructions. Before starting your mini-interviews, clarify specifics of the situation: age and gender of the client, setting of the interview, and other aspects essential to getting started.
3. Prepare a short questionnaire or a set of open-ended statements and ask your coached client to *complete them before your role play*. Their content should be appropriate to the topic of the role-playing situation.
4. Conduct a five-minute interview for each role-playing situation on the list, using content you learned from reading the client's responses on the structured writing. Record your interaction or have another person observe the interview, especially noting aspects related to

your response on the questionnaire or open-ended sentences. Start the interview by either the client or you summarizing the client's concern.

5. Use the responses on the questionnaire or open-ended sentences as the basis for your interaction.
6. Following each mini-interview, ask your coached client for feedback about the interaction. Then listen to the tape or talk to your observer and answer the questions in the "Focus Checklist," which follows the list of role plays.
7. Try some mini-interviews in which you practice appropriately integrating the techniques you learned in Parts 1 and 2 with use of questionnaires and open-ended sentences. You can create your own role plays or use any of the situations presented in the first two parts of this book. Use the "Summary Checklist" and record all appropriate data from these more integrated interviews. The "Summary Checklist" and its instructions follow the "Focus Checklist."

Role play

1. Your client is a teenager, an excellent student who seems very reluctant to share information with you, showing very shy, withdrawn behavior. To gain the needed information about the problem and because you believe the client may feel more comfortable writing, you devise ten open-ended sentences appropriate for this teenager, who is obviously unhappy, to complete. Have your client complete the sentences and then role-play from there.

2. Part of your job as executive director of the Youth Council is to interview adult volunteers who want to work with young people, to be sure that the volunteers are appropriate role models. Devise a short questionnaire that will help you initiate a screening interview with a volunteer. Your coached client will complete the questionnaire, and you will conduct the interview based on information given in the questionnaire.

Focus checklist

1. How did you use the responses to the questionnaire or the completed open-ended sentences to gain additional information about the client's thoughts, feelings, and actions?
2. Did you use information from the questionnaire or open-ended sentences to verify previous observations about the client?
3. Are the responses useful? Did the client understand what was asked and respond in an appropriate manner? If not, did you comment on the inappropriateness of the responses?
4. Did you clarify responses given on the questionnaire or open-ended sentences by further questioning or using other techniques?
5. Did you keep the interview future-directed toward counseling goals, even if some responses related to the past?

6. Are the questions or open-ended sentences appropriate for the purposes of the helping interview? Did the client understand what they meant? Were the questions/statements clear, focused, and written in language understandable by the client?

Summary checklist

How many of the following elements did you use, and in what ways did you use them during your more integrated mini-interviews?

1. Structuring the helping relationship
2. Focus on thoughts, feelings, actions, or TFA combinations
3. Listening for understanding and nonverbal aspects
4. Asking open questions
5. Asking closed questions
6. Systematic inquiry
7. Clarification
8. Reflection
9. Use of silence
10. Confrontation
11. Client concerns
12. Goals
13. Procedures
14. Evaluation
15. Termination and follow-up

Instructions

To assess your interview skills, make a worksheet similar to the Behavior Summary Checklist. In the *left column* on the checklist, make a transcript of key words and phrases or sentences used by you and your client during the mini-interview. Then, *for each* transcript response, put a check in columns that are relevant. You might check from one to several items for each response.

When you have completed the Behavior Summary Checklist, review it carefully. Look for patterns where you emphasized or ignored content. Then write a brief summary of the mini-interview outlining your strengths, limitations, and suggestions for improving skills.

References and recommended reading

Barker, C., & Lemle, R. (1984). The helping process in couples. *American Journal of Community Psychology, 12*(3), 321–336. These researchers use a self-report inventory to learn the personal and relationship history of the client as well as their experience in therapy or marriage enrichment training.

Cautela, J. R., Cautela, J., & Esonis, S. (1983). *Forms for behavior analysis with*

Summary Behavior Checklist

Key words and phrases of the Client (CL) and the Helper (H)	Focus or Emphasis			Structure	Nonverbal aspect	Open question	Closed question	Systematic inquiry	Clarification	Reflection	Confrontation	Use of silence	Client's concern	Goals	Procedures	Evaluation	Termination	Followup	Strategies
	T	F	A																

children. Champaign, IL: Research Press. This book contains a variety of tools for assessing children.

Clark-Lempers, D. S., Lempers, J. D., & Netusil, A. J. (1990). Family financial stress, parental support, and young adolescents' academic achievement and depressive symptoms. *Journal of Early Adolescence, 10*(1), 21–36. Researchers developed questionnaires to gather demographic data about parents and children in this study.

Corey, M. S. & Corey, G. (1987). *Groups: Process and practice* (3rd ed.). Pacific Grove, CA: Brooks/Cole. Chapter 11, "Groups for Adults," describes open-ended sentence-completion exercises.

Craighead, L. W. (1984). Sequencing of behavior therapy and pharmacotherapy for obesity. *Journal of Consulting and Clinical Psychology, 52*(2), 190–199. Craighead describes the use of an initial and follow-up self-report questionnaire for a group of obese clients.

Martin, D. G. (1983). *Counseling and therapy skills*. Prospect Heights, IL: Waveland Press. Martin discusses at length the use of audiotape in Chapter 8, "Observing Others," as a technique for beginning helpers.

Parker, W. M., Valley, M. M., & Geary, C. A. (1986). Acquiring cultural knowledge for counselors in training: A multifaceted approach. *Counselor Education and Supervision, 26*(1), 61–71. These authors use a structured questionnaire to collect data to assess knowledge of cultural awareness of counselors-in-training.

Woody, R. H., Hansen, J. C., & Rossberg, R. H. (1989). *Counseling psychology strategies and services*. Pacific Grove, CA: Brooks/Cole. Many uses for assessment tools are described in Chapter 8, "Clinical Assessment."

Charting Behavior and Recording Progress of Behavior Change

17

Key points

1. Charting behavior keeps track of specific aspects of behavior in a systematic way.
2. Charting behavior allows the helper and client to learn the frequency, conditions of occurrence, and direction (whether increasing or decreasing) of behavior.
3. Collecting baseline data enables the helper and client to determine the dimensions of problem behavior and to make a reasonable plan for change.
4. Charting behavior leads to accurate awareness and a clearer understanding of the situation.
5. Charting behavior shows what progress is occurring and puts into perspective the variation in the rate of progress.

CHARTING behavior is a systematic way of keeping track of specific behavior. The desired outcome of any helping process is behavior change—a difference in how the client thinks, feels, or acts. Often, change is slow or difficult to detect, and charting progress toward a goal enables both the client and helper to see positive direction and gives encouragement to continue with the helping program. If no progress is being made or if change is in a negative direction (if the problem is getting worse), another plan must be made. If, despite a dietitian's best instructions, for example, the charted weight of an obese client continues to increase or fails to decrease, it suggests that specific aspects of the situation must be examined or a different method of working with the overweight person must be tried.

- HELPER: This diet is obviously not working for you. Let's take a closer look at your situation and what your charts suggest.

Charting frequently provides specific data that can assist in making more effective plans for the future.

Charting behavior by the client or by someone near the client, such as a parent or teacher, is often a first step toward planning how to achieve behavior change. Collecting baseline data is important in establishing the severity and frequency of a problem. If a middle schooler (grades 6–8) gets into fights, how often do they occur—twice daily or once every three months? Is a fight a ten-word conversation or a physical battle requiring a trip to the emergency room? Do memory lapses of an elderly patient occur hourly or once weekly if medication is omitted? Acquiring such objective data will greatly assist a helper to understand the parameters of a problem. A frustrated spouse of an elderly person may easily exaggerate the frequency of time confusions. A father, trying to justify overly aggressive behavior with the "boys will be boys" coverup, may grossly underestimate the number and severity of his twelve-year-old son's hostile encounters.

- HELPER: Jan, I want you to write some things on paper every time you and someone else have a disagreement in the next week. Record when, where, and what actually happens and between which persons. Bring this outline with you next week to our appointment.

Dimensions of charting behavior

A helper must guide the client carefully in *what* as well as *how* to chart behavior. As seen in Chapter 2, on the TFA triangle, when one constructs a behavior blueprint, specific aspects of the client's thoughts, feelings, and actions can be seen. The behavioral characteristic to be charted must be clearly defined and understood by the client.

1. Behavioral specificity is essential in charting. One cannot, for example, chart *anxiety* but can chart *symptoms of anxiety* such as rapid breathing, sweating, stomach clenching, and feelings of panic.
2. Either frequency or intensity, or both, of a behavior can be recorded. The helper will want to discuss with the client whether to try to describe the *severity* of the behavior or just the *incidence* of symptoms. Will the client try to chart respiration rate during anxiety or simply record that rapid breathing occurred?
3. The helper must discuss with the client how the behavior is to be observed or identified. Deciding how behavior is to be determined is another important dimension of charting. Will the client have an observer who will give a cue when behavior is present, or will the client personally decide when behavior occurs? With feelings or thoughts, of course, only the client can determine what is present; with actions, others may be able to observe.
4. A behavior can be charted as part of a *process* or as a *product*. A client can record, for example, behavioral signs of anger or loss of control resulting in a "temper tantrum"; one is process and one is outcome, or product. Either or both may be important to the discussions of the helper and client.
5. To be successful, charting must be easy to do. Using a counter (e.g., a golf scorekeeper) to record specific behaviors as they occur allows immediate recording of how often, for example, a client says hello to an acquaintance in a day. A checklist placed on the desk can be a way to record use of time; a harried professional can gain valuable information for time-management changes by charting such daily interruptions as telephone calls and drop-in visitors. Or recording on a chart may take place once a day or once a week, such as a recovering alcoholic checking off alcohol-free days each evening. However done, the method must be clear to and convenient for the client, to make the charting work.

Uses of charting behavior

The technique of charting behavior is useful for several different reasons:

1. Charting provides baseline data. Recording frequency of desired or undesired thoughts, feelings, and actions provides an accurate awareness of specific behavior before there is any attempt to make changes; frequency includes what happens and when, where, and under what circumstances.

Frequently, just charting problem behavior leads to accurate insight about it. Further, this awareness can suggest specific things that must be changed to achieve goals and can help in setting realistic immediate and long-range goals. If baseline data indicate six angry outbursts a day, a goal of no outbursts in the next week is totally unrealistic. Reducing the number each day, or even each week, is a much more achievable goal in the early weeks of the relationship. Often, it is helpful if a client records the frequency of an action, as in the following example:

- HELPER: Looks like you spent about thirty minutes each night last week on homework, Pat. You said you made Ds on two tests and got two zeros in math. Based on that, what do you think might be a reasonable amount of time to put in on homework each night this week?

2. Charting enables the client and helper to see progress toward a goal, however slow that progress may be. If there is no (or extremely slow) progress, the helper and client can use this information to look for causes and alter procedures both during and outside of the interview.

- CLIENT: Last semester I had a 3.3 grade-point average, and now it's 3.46. The studying is paying off—I may get into medical school after all!

3. Charting shows both the ups and downs over a period of time (net gain or loss). Occasionally, an unwanted behavior may temporarily increase, causing the helper and client to feel discouraged, which may lead to even less progress. Charting enables both to see the overall picture, rather than just the most recent or the most dramatic segment of behavior. For example, a student may receive fewer and fewer detentions over a period of six weeks, then receive four detentions in one week and blow a good record, returning to only one detention again the next week. A chart that shows such an increase as an isolated peak in a declining series of incidents places the bad week in perspective.

- CLIENT: I thought we were making real progress getting Stephen to do his chores, but last week it seemed as if everything fell apart. When I look at the chart, though, I guess we really had two terrible days when he had friends over, and the rest were okay. It just seemed as if it took all week!

4. Charting shows specific aspects of behavior that may help point out specific problems.

- CLIENT: It seems as if I can stay within my budget when I am at home, but I hopelessly overspend if I go out with my buddies to a bar and restaurant. Maybe that's why my wife gets mad at me on the weekends when I go out.

5. Charting is simple and easy to do, provided the client is committed to doing it. The helper and client should discuss how often behavior will be recorded (once a day? once an hour? every time it occurs?) and where (on a wall chart above the bathroom scale? on a clicking device on which the client depresses a clicker every time the behavior occurs? on a pad of paper in the pocket or on the desk and used every fifteen minutes?).

- HELPER: Watch how I use this gadget that is usually worn on the wrist to record a golf score. I want you to press the button every time you get into an argument with your brother tomorrow.

Limitations of charting behavior

Charting behavior will be useful only if

1. The behavior can be broken down into specific, identifiable instances for counting. Something as nebulous as feeling angry would be more difficult to identify and record; "shouting at spouse" is more specific and recordable.
2. The client is willing to record *every* instance of the behavior in question.
3. The client is truthful.
4. A convenient method of charting can be developed for the client. If it's not convenient, it probably won't be used.

Charting behavior and the change cycle

Charting is a useful strategy in the first, second, and last stages of the behavior cycle. In the first stage, charting is often used to gather baseline data. What is the client's current behavior? Frequently, the client only has a vague notion of actual behavior ("I don't study a whole lot"; "Maybe I have two or three drinks a day"; "We seem to be fighting all the time"; "My boss is always getting on my case"). Thus, charting behavior enables both the client and helper to start with an accurate assessment of thoughts, feelings, and actions in relationship to the specific situation.

After baseline data have been gathered, the client and helper can set achievable, realistic goals. Rather than initially going on supposition, data gathering enables the helper and client to determine frequency and strength of the behavior to be encouraged or discouraged.

In the last stage of the behavior cycle, charting is also useful in keeping track of changes made and new behavior. Charting provides specific indications of the *rate* of change. The helper can note major advances as well as rough spots and assist the client in fine-tuning procedures and adapting them to individual differences. Finally, charting enables the client and helper to note the *amount* of change that has taken place. Accurate records of client changes can be useful to the helper in better understanding what works, for which clients, with which problems, in which situations.

Case study: Use of charting behavior

Mr. Simmons sought help to stop smoking. He is fifty years old, slightly overweight, and a principal of a large high school. His physician has warned him that he is a prime candidate for a heart attack and that he must change aspects of his life, beginning with cigarette smoking. In an attempt to model good behavior for their students, his faculty had just passed a no-smoking rule for the entire school building. Mr. Simmons was finding it difficult not to smoke during his ten-hour workday and felt guilty every time he shut his door to sneak a puff.

In his usual competitive manner, Mr. Simmons challenged himself to quit cold turkey but was cigarette-free for only two days before returning to cigarettes and admitting to himself that he needed help. Feeling somewhat foolish but determined to be honest, he related to the helper how he had tried complete withdrawal and even chewing tobacco, admitting that the latter method was neither satisfying nor socially acceptable (his wife refused to kiss him) nor consistent with his self-image as a professional educator.

In answer to the helper's question of how much he smoked, he sheepishly admitted he didn't know. "I buy only a pack of cigarettes a day, but I bum a smoke from every smoker I know. Because I'm the principal, naturally the staff gives me one when I ask—especially if I let them sneak a smoke with me in my office." The helper determined that it was necessary to collect data and define the problem before goal setting and strategy planning could begin. She asked Mr. Simmons to note the time, place, and situation for every cigarette he smoked during the week before his next interview.

Based on the data he brought, Mr. Simmons and the helper could determine time, frequency, and situations associated with his smoking. They then made a plan to help him begin to reduce smoking in and out of school. Charting during subsequent weeks allowed Mr. Simmons to see in graphic form that he was, in fact, making substantial progress. And the helper could remind him that, even though consumption went up the week of the three bomb threats, the teacher strike, and the budget hearings, overall he was making excellent, if slow, consistent progress toward his short-term goal of cutting down from thirty-five to ten cigarettes each day. Ultimately, the helper hoped to help Mr. Simmons achieve his long-term goal of no cigarettes, which he would record as smoke-free days on a calendar. The helper would enable Mr. Simmons to shift finally from recording a negative behavior (smoking) to a positive one (tobacco-free days).

Analysis

By having Mr. Simmons chart his cigarette smoking, the helper can

1. Acquire baseline data that helper and client can use to know specific data related to the client's concern.
2. Plan realistic goals and effective strategies for change, based on concrete details associated with the client's concern.

3. Involve the client directly in his own treatment plan by making him aware of every time he practiced the behavior (cigarette smoking) he was trying to reduce.
4. Help the client see continuous progress, even when temporary lapses occur.
5. See when and where lapses did occur and help the client plan against them.
6. Gain tangible weekly feedback on whether counseling sessions were helping the client.
7. Help the client move from charting negative behavior (smoking) to more self-reinforcing charting of positive behavior (tobacco-free days recorded on a calendar).

Practice

1. Find a person who agrees to help you learn how to use the strategy of charting behavior. Assume that you and your coached client have previously conceptualized the client's concerns.
2. Select one situation from the list of role-plays, which follows these instructions. Before starting your mini-interviews, clarify specifics of the situation: age and gender of the client, setting of the interview, and other aspects essential to getting started.
3. Conduct a ten-minute interview for each role-playing situation. Record your interaction or have another person observe the interview, especially noting aspects related to giving instructions for writing and discussing the client's chart. Start the interview by either the client or you summarizing the client's concern.
4. Use the client's chart to learn about thoughts, feelings, and actions related to the role play. Discuss with the client those thoughts, feelings, and actions and how you see them relating to the concern you have discussed.
5. Following each mini-interview, ask your coached client for feedback about the interaction. Then listen to the tape or talk to your observer and answer the questions in the "Focus Checklist," which follows the list of role-playing situations.
6. Try some mini-interviews in which you practice appropriately integrating the techniques you learned in Parts 1 and 2 with use of charting behavior. You can create your own role-playing situations or use any of the situations presented in the first two parts of this book. Use the "Summary Checklist" and record all appropriate data from these more integrated interviews. The "Summary Checklist" and its instructions follow the "Focus Checklist."

Role play

1. Your client complains of being continually harassed by co-workers, including sexual harassment. You want to know what the client considers

Behavior	Monday	Tuesday	Wednesday	Thursday	Friday
Touching Jokes Criticism Other harassment					

By "other harassment" I mean:

1. ______________________________
2. ______________________________
3. ______________________________
4. ______________________________
5. ______________________________

Sample chart for role-playing situation 1. The client is to make a mark in the appropriate column each time a particular behavior occurs.

harassment, how often this harassment is occurring, and just what kind of behavior is involved. As personnel manager, you are aware that a particularly explicit description of this behavior is necessary because of legal implications for your company if the charges are true. Explain to the client what kind of chart you wish to have prepared, how to keep the chart, and what use you will make of the data. The chart might look something like that above, with a mark for each time the behavior occurs.

2. Choose a behavior of your own and chart its frequency and strength. Pay particular attention to devising a method to chart easily and accurately. Note difficulties you have in specifying or charting behavior of your own and think how this difficulty may relate to asking clients to chart their behavior. Discuss what you have learned with another helper-in-training and come up with changes in your method of charting to make data gathering easier or more accurate.

3. Your client, a twelfth grader trying unsuccessfully to pass advanced math, has brought you a chart that you requested of study hours and interrupting factors. Discuss the behavior chart with your client and use it to set goals and plan strategies to meet the long-range goals of passing the class. (Ask your coached client to prepare the necessary chart before the interview begins.)

Focus checklist

1. What useful information did you gain about thoughts, feelings, and actions from what the client had charted?
2. Did you give the client clear instructions about how to keep a chart?
3. How did you relate what was in the chart to the client's present concern?

4. Did you explain to the client whether to continue the chart, and if so, what modifications to make in the recording procedure?
5. Was the charted behavior used to gather baseline data or to monitor changes in behavior?
6. Did you and the client discuss a method for making a chart for recording the client's behavior? Was it clear to the client when, how, and what to record?
7. How did you lead the client to acquire insights into and draw conclusions about the data?

Summary checklist

How many of the following elements did you use, and in what ways did you use them during your more integrated mini-interviews?

1. Structuring the helping relationship
2. Focus on thoughts, feelings, actions, or TFA combinations
3. Listening for understanding and nonverbal aspects
4. Asking open questions
5. Asking closed questions
6. Systematic inquiry
7. Clarification
8. Reflection
9. Use of silence
10. Confrontation
11. Client concerns
12. Goals
13. Procedures
14. Evaluation
15. Termination and follow-up

Instructions

To assess your interview skills, make a worksheet similar to the Behavior Summary Checklist. In the *left column* on the checklist, make a transcript of key words and phrases or sentences used by you and your client during the mini-interview. Then, *for each* transcript response, put a check in columns that are relevant. You might check from one to several items for each response.

When you have completed the Behavior Summary Checklist, review it carefully. Look for patterns where you emphasized or ignored content. Then write a brief summary of the mini-interview outlining your strengths, limitations, and suggestions for improving skills.

References and recommended reading

Amundson, N. E. (1987). A visual means of organizing career information. *Journal of Employment Counseling, 24*(1), 2–6. Amundson shows a method of organizing information visually for clients in career counseling.

Summary Behavior Checklist

Key words and phrases of the Client (CL) and the Helper (H)

T F A

Focus or Emphasis	
T	
F	
A	
Structure	
Nonverbal aspect	
Open question	
Closed question	
Systematic inquiry	
Clarification	
Reflection	
Confrontation	
Use of silence	
Client's concern	
Goals	
Procedures	
Evaluation	
Termination	
Followup	
Strategies	

Bellack, A. S., & Hersen, M. (1988). *Behavioral assessment: A practical handbook* (3rd ed.). New York: Pergamon Press. The authors discuss a wide variety of assessment means.

Dixon, D. N., & Glover, J. A. (1984). *Counseling: A problem-solving approach*. New York: Wiley. Dixon and Glover give ideas of how to assess client progress in solving problems.

Konczak, L. J., & Johnson, C. M. (1983). Reducing inappropriate verbalizations in a sheltered workshop through differential reinforcement of other behavior. *Education and Training of the Mentally Retarded, 18*(2), 120–124. The authors provide a clear example of charting client behavior to assess progress toward a goal.

Krumboltz, J. D., & Krumboltz, H. B. (1972). *Changing children's behavior*. Englewood Cliffs, NJ: Prentice-Hall. These authors suggest and explain graphing change and using checklists to see progress.

Rohsenow, D. J., & Smith, R. E. (1982). Irrational beliefs as predictors of negative affective states. *Motivation and Emotion, 6*(4), 299–314. An account of irrational beliefs related to dysfunctional emotional response, the article includes descriptions of clients who kept daily records of anxiety, anger, unhappiness, and drinking behaviors for seven months.

Contracting

18

Key points

1. A helping contract specifies behavior to be changed and the positive or negative consequences of accomplishing (or not accomplishing) the change.
2. A helping contract can be between two or more parties, for example, teacher and student; spouses; parents and child; employer and employee; family members; juvenile offender, probation officer, and parent or guardian.
3. A client sometimes makes a personal contract to change behavior and receive a self-arranged reward or reward from someone else.
4. A helping contract should be specific, concrete, achievable, positive, and future-oriented.

THE idea of contracting is implicit in all helping relationships. With the aid of the helper, the client contracts (or makes an agreement) to set specific goals and plan a strategy to meet those goals. The set of strategies (the plan for change) becomes the contract, and the achieved goals are the client's reward for fulfilling the terms of the contract. A good contract has four aspects:

Specificity

1. A written or verbal contract is an explicit plan whereby the client agrees to meet certain conditions and/or perform actions, which will be rewarded in a way specified by the contract. A specified behavior or set of behaviors is rewarded in an agreed-on manner. Failure to engage in the specified behavior receives negative consequences such as denial of a privilege or punishment. One typically thinks of a contract as involving at least two parties: parent and child, supervisor and employee, nurse and patient, husband and wife. Such may be the case with a helping contract, devised between two or more parties.

- HELPER: Do you understand, Jonathan, that if you call your sister names, you will spend thirty minutes in your room by yourself?

Self-contracting

2. In the helping process, it is possible for there to be only one party to a contract: The client may establish the behavior desired and be self-rewarding for achieving the behavior. The helper assists the client to achieve stimulus control, to control his or her own behavior in a given situation, rather than being subject to external controls.

- CLIENT: I will not shoot baskets until I've finished my math homework each day.

Delayed gratification

3. Although delayed gratification (taking a reward after a task is completed) is the norm for many persons, clients used to acting on impulse can be

taught to reward themselves after performing a task or avoiding a particular behavior.

- CLIENT: I've agreed with my diet counselor that every time I lose five pounds I will buy myself a small article of clothing, and at every ten-pound loss I will buy something to show my physique, like a new swim suit.

Personalization

4. Sometimes a reward that is implicit in a system is not seen as personally meaningful to a client. If a given behavior is not seen as personally meaningful, the odds of the client doing it are very low. With professors in a university, research and publication are meaningful; in business, sales personnel give customers personal attention because it is directly linked to a dollar bonus. Teachers consider good grades and the joy of learning sufficient reward for diligent effort, but for many students, these concepts are so abstract they are meaningless as reinforcers of behavior. A school counselor can personalize the school's reward system so that it has more meaning in a student's life.

- HELPER: If you make a *C* in Business English, type fifty words a minute, and have fewer than ten absences this semester, you can enter our job-training program in the second semester, work in an office two hours a day, and earn minimum wage while you get school credit for the work experience.

This kind of specific contract makes much more immediate sense to an indifferent student than does the student handbook's statement that good grades and typing skill will help you get a job.

Similarly, positive rewards are generally more effective than negative ones. The most desirable contract specifies a positive outcome for a client, rather than avoidance of a negative outcome. While avoidance of negative outcomes is probably a part of any contract, emphasis should be on the positive goals of behavior, rather than the negative (see Chapter 17).

Uses of contracting

The technique of contracting in the helping relationship is very useful for several reasons.

1. A contract pinpoints very specifically the behavior to be encouraged or erased.

- CLIENT: After thirty minutes of dry-land exercise and swimming laps, I will reward myself by drinking a cup of coffee and reading the morning newspaper at my favorite donut shop.

2. A contract details in an if-then construction the immediate reward for a stated action. A specific reward will reinforce the behavior and increase the

likelihood of repetition of that behavior. A reward that is a natural consequence, such as buying new clothes following weight loss, is likely to be most effective. Or, if extinguishing the behavior is the goal, the consequence is specified in the contract and is most effective if immediate and a natural consequence. A contract for a shy student, for example, might state:

- HELPER: If you raise your hand in class and answer a question or volunteer a comment on the topic, you will receive a daily grade of *A* for classwork.

3. A helping contract may allow persons to negotiate interaction. This may result in a compromise about an acceptable behavior or may involve one person rewarding or punishing another for certain actions. In either case, all parties to the contract know what is expected of each and what consequence or rewards are to be enacted.

- HELPER: If June takes the management course and works on application of principles of communication, Cynthia agrees not to intervene in day-to-day management of June's department. Cynthia and June will meet biweekly to review June's progress. Cynthia will not issue orders to counter June's directions, nor will she contradict, reprimand, or otherwise correct June in front of June's co-workers.

4. A written contract is concrete, something that can be kept for future reference. It is a promise on paper, a gift of hope that if the client changes in certain ways, then certain desirable things will happen or certain undesirable things won't happen. It is tangible evidence that the helping process can work for this particular client according to a specified procedure. As a textbook is a comfort and source of security to a graduate student, a simple, explicit contract can be a promise of change for a client in a helping relationship. The client's reaction to a contract may go something like:

- CLIENT: If I study, *really study,* for two hours a day, my parents said they'll stay off my back about grades. And I'll bet if I do, I might even be able to pass English this term. Two hours a night will still leave me about four hours between school and bedtime to do what I want. I think I can do that—I'll try it, anyway.

5. A contract is future-directed in that it focuses on behavior to be achieved or practiced and on rewards to be given for that behavior. It directs the attention of the helper and client to what will happen.

- CLIENT: Beginning tomorrow morning, I will allow myself no novel-reading (or TV) until I've put in one hour of housework. I'm not going to worry about what has not been done in the last three months; I'm just not going to pick up a novel until I've worked for at least one hour. I've been wasting my time buried in novels when I needed to do other things around the house.

6. A special use of contracting is often with suicidal clients. Beyond the scope of the neophyte helper, presumably the reader of this text, this use is often employed by therapists who work with suicidal clients. Typically, such a

contract requires that a client contact the helper when suicidal thoughts predominate and threaten and that the helper be available when needed by the client in a crisis. The client promises not to leave; the helper promises to be present.

Limitations of contracting

The contracting strategy in the helping relationship has definite limits.

1. A contract speaks only to a certain behavior or set of behaviors. Because a contract is narrow and specific, it can be effective, but it can also ignore other areas that need equal attention. The helper must discuss with the client that a particular contract is only one aspect of the helping process and that several contracts for specific behaviors may be required over the course of the helping process.
2. A contract frequently involves agreements between parties. One party may be less willing to commit to the contracted behavior than the other. If one party fails to participate as the contract states, this will obviously affect the participation and success of the other party. Mutual trust between contractors is necessary if a multiparty contract is to be successful—at least, there must be trust on the specific areas covered by the contract.
3. Factors outside the helper's and client's control may affect the outcome of a contract. If a student makes a contract with the assistant principal and another person assumes the role of the assistant principal, the original contract might not be carried out.
4. Frequently, contracts are written too broadly for success to be defined. Statements such as "John will not challenge his parents' authority" is a statement far too broad for successful evaluation of completion. It must be broken down into what, specifically, is meant by *not* challenging parental authority.
5. Sometimes contracts specify too many behaviors, and the client cannot deal with them all at once. A good contract is realistic and achievable; more behavior change can be negotiated in future contracts.
6. If the contract specifies a negative behavior to be reduced or extinguished, success is unlikely unless the consequence is natural. A contract reading "I will run around the block one time whenever I bite my fingernails" is unlikely to be successful because the consequence probably can't—and won't—be carried out. Positive-consequence contracts are much more desirable than negative-consequence documents.

Contracting and the change cycle

Contracting is important in goal-setting because the contract itself specifies at least a short-range behavior change goal. Specifics of behavior are outlined in the contract, constituting a series of short-range action goals.

Contracting is especially used in the procedure phase of the behavior cycle when the client is actively engaged in taking specific action to change behavior. Contracts can be specifically targeted toward changes in the client's thinking, feeling, and acting, as well as toward the situation and others who may be involved. As the contract is monitored, clues can emerge that provide specific data the helper can use to assist the client in making specific changes to increase the probability of progress.

Case study: Use of contracting

Samantha felt as if she was under surveillance all the time. The helper listened to Samantha's laments and got her to specify ways in which she was feeling harassed by Mrs. Bryant, the assistant principal. From previous conversations, the helper knew that the assistant principal did indeed regard Samantha as an extremely rebellious young woman who was a strong leader in influencing other students to defy authority, especially as represented by the assistant principal. The helper also determined that Samantha's classroom behavior was, if not exemplary, at least acceptable and that she was earning passing grades, except for times when she was suspended from school for misbehavior.

The helper discerned from her conversations with Samantha, however, that the young woman seemed to crave praise and particularly felt anger because she was unable to elicit positive recognition from Mrs. Bryant. Well aware that Samantha had given Mrs. Bryant little, if any, opportunity for positive recognition in the past, the helper nonetheless proposed that Mrs. Bryant reward Samantha for good behavior when it occurred.

After the helper had received positive but dubious agreement from the assistant principal, they drew up a contract that read as follows:

For the next month, Samantha will

1. Be in class when the tardy bell rings.
2. Avoid pushing, tripping, shouting, cursing, or otherwise intimidating students and adults in the building.
3. Be in her assigned place in the school at all times.
4. Smoke only off of school property.

For every day that Samantha met the above conditions, she would meet Mrs. Bryant at 8 AM the following day and put up the flag in front of the school, a job of some distinction with other students.

Both signed the contract, with the helper as witness, and kept a signed copy. After one week, the helper asked Samantha how it was going. Samantha replied, "At first, it was hard. I was tardy to two classes on Monday, and I didn't get to put up the flag. But Mrs. Bryant actually said some nice things to me about the other stuff I did right. The next day I did okay, and Wednesday I got to put up the flag, with everybody watching. Thursday I did, too, but then I messed up and Friday somebody else put up my flag. Next week I bet I get to put it up every day!"

In response to the same question, Mrs. Bryant said, "When I started *working with* her, I realized Samantha was really doing most things right. I felt that she was challenging my authority and that I had to face her down. It was difficult for a while, but I'm really trying to be positive to her. And I must admit, there've been no discipline referrals on her this week, and it makes me feel good to see how proud she is when she puts up the flag. Next week I'm going to call her aunt with whom she lives and tell her how well Samantha is doing. Goodness knows I've made enough negative phone calls to her; she'll probably faint when I tell her how well Samantha is doing."

Analysis

Using the contract method in this helping relationship enabled the helper to

1. *Assist both parties* to negotiate behavior acceptable to both. Each party agreed to certain behavior, and each party received a reward. No one lost in the interaction, and each party won the reward she wanted.
2. *Focus attention* on desirable behavior in the future, rather than getting stuck in past behavior.
3. *Clarify* exactly what was expected of each. For Samantha, this involved abiding by school rules, as well as personalizing which rules she needed to heed. For Mrs. Bryant, the agreement was for positive recognition and a definite reward—raising the flag—for good behavior.
4. *Emphasize action,* not feelings. According to the contract, Mrs. Bryant and Samantha would behave in certain ways. They were still free to totally dislike one another if they chose. Because actions, not feelings, were causing the overt problem, the helper concentrated on changing those actions. Samantha's comment about Mrs. Bryant being friendly to her would give hope that feelings, too, might someday change. But the contract focused on actions. Mrs. Bryant's comment that she realized that she felt Samantha challenged her authority indicated there was a chance her thinking about Samantha might change in the future, too—but again, this was not required under the terms of the contract.

Practice

1. Find a person who agrees to help you learn how to use the strategy of contracting. Assume that you and your coached client have previously conceptualized the client's concerns.
2. Select one situation from the list of role-plays that follows these instructions. Before starting your mini-interviews, clarify specifics of the situation: age and gender of the client, setting of the interview, and other aspects essential to getting started.
3. Conduct a ten-minute interview for each role-playing situation. Record your interaction or have another person observe the interview, especially noting aspects related to giving instructions about the

contract and your discussion of contract results. To start the interview, either the client or you should summarize the client's concern.

4. Use the client's experience with the contract to discuss progress in changing thoughts, feelings, and actions.
5. Following each mini-interview, ask your coached client for feedback about the interaction. Then listen to the tape or talk to your observer and answer the questions in the "Focus Checklist," which follows the list of role-playing situations.
6. Try some mini-interviews in which you practice appropriately integrating the techniques you learned in Parts 1–3 with use of contracting. You can create your own role-playing situations or use any of the situations presented in the first two parts of this book. Use the "Summary Checklist" and record all appropriate data from these more integrated interviews. The "Summary Checklist" and its instructions follow the "Focus Checklist."

Role play

1. Your client is a forty-five-year-old woman who wants to spend more time away from the job. You want to devise a contract to reward away-from-work time and to specify conditions under which the client will leave the workplace *and* not take work home. Discuss with the client what behaviors this will entail and draw up a simple contract with self-rewards.

2. Discuss with a client a contract (which you've already prepared) to change study behavior that was unsuccessful, probably because it was unrealistic. Point out unrealistic expectations with regard to study behavior (make up some; be creative!) and discuss what points might be included in a realistic, achievable contract on the same topic.

3. Your client is a sixteen-year-old boy who "borrowed" a stranger's car for a short ride with his friends. He realizes the error, is contrite, and has a supportive single parent. With this client and his mother, draw up a contract that includes minimal contact with the court system and with you, the probation officer.

Focus checklist

1. How did you explain to the client the purpose and goals of contracting?
2. Did you cover the concepts of delayed gratification and the principles of behavior change and behavior reinforcement?
3. When the contract was written, was it designed to contain logical consequences, a clear plan, and specificity of client actions and rewards to be attained?
4. Is time for evaluating progress and for completing the contract either built into the contract or explained to the client?
5. Are all parts of the contract clear and agreeable to the client?

6. Are contracted behaviors achievable for the client, and does the client perceive them as such?

Summary checklist

How many of the following elements did you use, and in what ways did you use them during your more integrated mini-interviews?

1. Structuring the helping relationship
2. Focus on thoughts, feelings, actions, or TFA combinations
3. Listening for understanding and nonverbal aspects
4. Asking open questions
5. Asking closed questions
6. Systematic inquiry
7. Clarification
8. Reflection
9. Use of silence
10. Confrontation
11. Client concerns
12. Goals
13. Procedures
14. Evaluation
15. Termination and follow-up

Instructions

To assess your interview skills, make a worksheet similar to the Behavior Summary Checklist. In the *left column* on the checklist, make a transcript of key words and phrases or sentences used by you and your client during the mini-interview. Then, *for each* transcript response, put a check in columns that are relevant. You might check from one to several items for each response.

When you have completed the Behavior Summary Checklist, review it carefully. Look for patterns where you emphasized or ignored content. Then write a brief summary of the mini-interview outlining your strengths, limitations, and suggestions for improving skills.

References and recommended reading

Cohen, B., & Sordo, I. (1984). Using reality therapy with adult offenders. *Journal of Offender Counseling Services and Rehabilitation*, *8*(3), 25–39. The authors describe a reality therapy model and explain their use of contracting for behavior change.

Corey, G., & Corey, M.S. (1982). *Groups: Process and practice* (2nd ed.). Pacific Grove, CA: Brooks/Cole. The authors discuss contracts combined with homework assignments as a way of achieving personal goals in counseling.

Hackney, H., & Cormier, L. S. (1979). *Counseling strategies and objectives* (2nd ed.).

Summary Behavior Checklist

Key words and phrases of the Client (CL) and the Helper (H)

T F A

Focus or Emphasis	
T	
F	
A	
Structure	
Nonverbal aspect	
Open question	
Closed question	
Systematic inquiry	
Clarification	
Reflection	
Confrontation	
Use of silence	
Client's concern	
Goals	
Procedures	
Evaluation	
Termination	
Followup	
Strategies	

Englewood Cliffs, NJ: Prentice-Hall. The section on self-management and self-monitoring includes a discussion of self-contracting.

Homme, L. E. (1970). *How to use contingency contracting in the classroom*. Champaign, IL: Research Press. This book gives a good overview of all essential components of contingency contracting.

Positive Self-Talk

19

Key points

1. Positive self-talk teaches a client to say good rather than bad things about self.
2. A client who thinks and says positive things tends to change thoughts, feelings, and actions about self.
3. Positive self-talk bridges the time between client–helper appointments: The client learns to replay internally some of the things heard from the helper in their meetings.
4. A helper can teach a client to recognize self-defeating thoughts.
5. A client can learn to replace self-defeating thoughts with positive messages and to stop negative thoughts.
6. A client can "psych up" through positive self-talk and learn to avoid putting down self through negative ones.
7. Positive imagery can help a client see self as successful in an impending situation.

MUCH of the forward motion toward behavior change develops during the helper–client interaction as the two set goals and strategies and assess progress. Through the trust and empathy already established, the helper's positive regard for the client, as well as encouraging words and recognition of accomplishment, creates an atmosphere of confidence and expectation of success for the client.

Part of the helper's goal is to enable the client to envision success sufficiently so that the feelings of confidence carry over until the next client–helper meeting. An important focus for the helping relationship is to assist the client to do things that will be perceived as successful—certainly by the client and hopefully by others who will give positive reinforcement to the client.

Positive self-talk is a technique that can help bridge the time between helper–client appointments. Just as the helper gives positive messages to the client, so can the client learn to give the same kinds of mental messages to self: positive self-talk, conversing with the self mentally.

Most of us are all too familiar with negative self-talk:

- "I just know I'm going to fail the test; I should have studied all night."
- "I'll never get a date; I'm too ugly."
- "My boss never says yes to my ideas."
- "I won't possibly get the promotion."
- "I've never been good at . . ."
- "Nobody would ever want to be my partner."

Negative self-talk, with its putdown-type messages, has been learned through the years by the client; however, the client can, with the assistance of the helper, learn the activity of positive self-talk with its uplifting effect. This process is called *cognitive restructuring*. A client can learn to replace negative thoughts with positive ones, to substitute the positive "I'm eating the right foods and exercising; think I'll get a carrot" for the negative, self-defeating "I'll never lose weight; might as well have a hot fudge sundae." This kind of auditory self-talk—when the client actually formulates the words in the mind—sends a message of positive, self-enhancing behavior. With proper training, the client can learn to think like this:

- "I studied, so I can do well on the test, or at least as well as most of the class."
- "Maybe I'll get a date; I've got a good tan, and I'm feeling pretty good right now."
- "If I develop this plan and show its cost effectiveness, I'll bet my boss gets really excited about it."
- "I'm really the best-qualified person for the job."
- "I'm going to ask Pat to be my partner."

The helper can assist the client to learn to turn off negative thoughts, but first the client must learn to recognize a negative, disabling thought. A client who is depressed, generally unsuccessful, and anticipating failure will likely not even recognize negativism in thinking.

By using self-reporting techniques—perhaps open-ended questions about the thought process (see Chapter 16), a diary of thoughts and feelings, or an open-ended questionnaire eliciting feelings about specific situations—the helper can aid the client in identifying typical negative thoughts or thought patterns.

- HELPER: I want you to answer these questions about what you think and feel when you find yourself facing certain situations.

Sometimes a client will be pervasively negative and self-defeating, seeming not ever to be positive in self-talk and expectations.

- CLIENT: What do you mean, what am I good at? I'm good at nothing! I'm a total failure.

It is important that the helper assist this client to identify specifically negative thoughts and feelings tied to particular persons, actions, or situations, rather than accepting globally negative thoughts. This narrowing process will occur partly as the helper and client conceptualize client concerns.

Other clients will be negative relative only to a particular situation. Such clients should be asked to recall specifically what was thought and felt at a particular time with regard to an incident, a particular person, or a situation so that the helper will have specific information with which to work.

- CLIENT: Whenever I get up to speak before a group, I get shaky in the knees, my voice trembles, and I really have to go to the bathroom. One time I got so nervous that I actually had to run from the room; now every time I know I have to speak, I remember that. So I try to avoid ever having to speak in groups. It's beginning to hurt me at work, though. I always try to get somebody else to be spokesperson, and my associates are beginning to get credit for my work.

Some clients have excess thoughts about good ideas. A young woman who, for example, visualizes herself so thin that she thinks herself into an eating disorder or into exercising to the point of damaging her body has turned positive thoughts into negative actions. Working with persons with eating disorders requires extreme skill on the part of the helper, but even the neophyte helper must be aware of symptoms and ready to refer an eating-disordered client for further lifesaving assistance.

Uses of positive self-talk[1]

The technique of positive self-talk can be used in several ways in a counseling setting:

1. Positive self-talk can be used to change thoughts, feelings, and actions. Thoughts and feelings influence action. Positive self-talk changes the way a client thinks and feels about self or performance or capabilities. Thinking positive thoughts creates a positive feeling about personal capabilities and responses, which in turn creates a much better chance for successful action. Clichés recognize this fact:

- Clothes make the man.
- You can if you think you can.

Norman Vincent Peale's *The Power of Positive Thinking* and Dale Carnegie's enormously successful classes in positive thinking illustrate the popularity and power of positive thought.

2. Learning to use the positive if-thens can give the client a feeling of confidence and an expectation of success. Frequently, before the client can learn to do this, however, helpers must assist their clients to recognize negative, self-defeating thoughts. A client might be asked to recall thoughts that came to mind before a tension-producing event such as a job interview or romantic encounter. The client may recollect self-talk—what was said mentally—along the lines of:

- "What if I'm late for the interview?"
- "What if I can't find a parking place?"
- "What if I trip going into the office, my voice squeaks, my hands sweat, I drop my portfolio and my papers fly out, I have to use the bathroom . . .?"

The what-ifs are generally very effective negative self-talk, preventing the person from considering positive, constructive, self-enhancing behavior.

To eliminate or change negative self-talk, the client must first learn to recognize it and then replace it with positive self-talk. *What-ifs*—negative

[1]At a basic level, positive self-talk can be used by a beginning helper. More advanced training and supervision will be required to use some of the systems described in the recommended reading for this chapter.

thoughts about what might happen—can be made specific and concrete either verbally or by writing them down on a list.

- HELPER: What I'd like you to do this week is to write down the negative thoughts you have [about not being able to do school work; about your child's behavior; about your personal feelings of inadequacy]. Keep a pocket notebook and write these down immediately or as soon after they happen as possible.

What-ifs *can* become positive. Proper prior planning and some intelligent anticipation of rough spots can turn bad what-ifs into positive if-thens, with the client creating a positive scenario instead of a negative one—stopping short of arrogance, of course.

- CLIENT: What if they offer me a job on the spot? I'd have to think about that. In any case, I'll bet I'll meet some neat new people, and I've always wanted to see the inside of that operation. I'll wear my new suit—I know I look good in that. And I'll take a taxi; that way I won't need to worry about a parking place. If they take me to lunch, then I'll know they're really interested in me.

3. Positive self-talk can systematically replace a negative message with a positive one. Some negative messages, of course, are not instantly replaceable: "I'm fat" can't become "I'm thin" just in the mind. But "I'm fat" can be immediately turned into:

- CLIENT: I'm on a diet, and I'm getting thinner. And I look better today than I did yesterday. And I'll make even more progress tomorrow.

This positive self-talk can translate into effective action: sticking to a diet.

4. The client can learn to use positive self-talk to "psych up" for dealing with a particular situation—reviewing strengths and strategies and anticipating how good performance is going to be.

- CLIENT: I've been to computer school, and I know this language inside and out. I know how the system operates. These malfunctions are temporary. I can debug this program and make it work for us.

5. Positive self-talk can stimulate images of success. People who are successful often say they image success: They actually visualize themselves completing a task competently and accepting congratulations or rewards for a job well done. The shy researcher who must deliver a paper to an international colloquium may be helped by developing a mental picture of standing on a podium looking poised and self-confident and delivering comments in a forceful, clear voice. The picture will *not* include what-ifs about losing a page of text, breaking glasses, knocking over the microphone, or tripping while ascending the steps to the speaker's platform. Of course, it will help if the client can examine the physical arrangements *before* going to the podium, see that notes are arranged so they won't get out of order, and so on; such

commonsense preparation enhances the probability of success or takes much of the potential worry out of the situation.

- CLIENT: I've finished speaking and they're applauding. Five people have their hands up to ask questions or to compliment my paper. The chair of the symposium is asking me to submit my work for publication, and a department head from a major university wants to have lunch tomorrow to discuss job prospects.

Although such thoughts are fantasy, *they should be potentially achievable*. The delivery may go just as pictured, and feelings and thoughts that project confidence will surely make a successful experience much more likely to occur.

Limitations of positive self-talk

Positive self-talk may sound too good to be true, and there are, indeed, limitations to its effectiveness:

1. Positive thoughts must be based on reality. Positive thoughts about bluegrass music do not make one a banjo picker—but believing one can learn surely enhances the effect of an hour's daily practice.
2. The client must not be lulled into false security or overconfidence. Thought and feeling must be followed by positive action.
3. The client must have a high probability of success in performing the task being visualized. If a client lacks the skills or personal characteristics necessary for a desired job, the helper must either assist the client to acquire those necessary traits or help the client weigh alternatives about accepting the job. One must think *accurate* positive thoughts to achieve success.
4. The client—and the helper—must believe the positive self-talk. Insincere or inaccurate feedback from the helper can result in unwarranted confidence, with resulting probable failure and reduced confidence in the helping process or in the helper.
5. All negative thoughts are not self-defeating. A helper does not intend to turn a client into an unrealistic Pollyanna who does not recognize negative aspects of reality.

Positive self-talk and the change cycle

The strategy is used most often at the goal-setting and the procedure stages of the behavior cycle. Visualizing success is a good way to help a client define specific goals: The image becomes the goal, a dream to be achieved.[2] In selecting procedures for changing behavior, either the client or helper may

[2]Thanks to James S. Vaught, Assistant Superintendent, Wythe County (Va.) Schools, for clarifying the role of dreams in counseling. As he said to the 1991 graduates, "Hang on to your dreams; they'll get you through."

recognize how negative self-defeating thoughts are hindering the client's behavior. They work together to identify *what* the client thinks as well as when and where. Positive self-talk begins with an awareness of negative, self--defeating thoughts. As the client learns to recognize negative thoughts and replace them with positive self-talk, changes will occur in not only thoughts but also actions and feelings. Learning to do positive self-talk can be among the most effective strategies a helper can teach a client for long-term good mental health.

Case study: Using positive self-talk

The personnel manager was very surprised to see Bob Jones appear at her desk, twisting safety glasses in his hand and very ill at ease. "How can I help you, Mr. Jones?" she inquired.

"It's this quality-circle thing. I'm really worried about it."

Knowing the man to be a top quality machinist, conscientious far beyond requirement, loyal to the plant for his twenty-odd years of employment there, noted for avoiding departmental infighting and tending to his own business, she was curious to hear more. "You don't think the idea will work?"

"No, ma'am, it's not that. I think it's a fine idea, and I know management will learn if they listen more to us line workers. I am glad I'm in the group. But I was elected to be the leader and to talk for our group. I just don't think I can do that. I tried to tell my buddies because at first I thought they were just kidding with me because I'm the oldest. But they really want me to be the leader. They say they always come to me, anyway."

"Come to you?"

"Yeah, you know, I been here a long time, and I like to just go off and listen to a younger guy at lunchtime that's maybe, you know, having a little home trouble, or not getting along too good with his boss, or having trouble learning a new procedure, or something like that. Usually I just listen or show him a little thing, and he goes on and things work better. But that's not being a leader, and the men say I'm already the leader."

"Yes, I know from what you've said before that you have a reputation as the person who makes the machine shop run well with people getting along in there. What worries you about the quality-circle group?"

"Well, there'll be a whole group of people, not just one or two, and we'll all be sitting around a table or something, like in an office, not the shop. What if I don't know the right words to say? What if nobody talks, and there's nothing to lead, and we just all sit quiet? What if I go to talk to the bosses and forget what to say, or dress wrong, or use bad grammar and sound ignorant to them?"

"You're bothered mainly by having to represent the group and speak for the other members."

"Yeah, that's it exactly. My wife says I talk all the time at home, and that if I can pray out loud in church, I can do this. But I don't know; the Lord may not give me words about work stuff like he does in church," he joked half-heartedly.

The personnel manager reminded him that leaders would receive training before they conducted their own quality-circle meeting, so he wouldn't have to think up everything to say each week all by himself. Promising to think to himself about the times he'd spoken well before groups in the past, he agreed to attend the leaders' seminars and reconsider his elected leadership role. With her help, he was able to think of several times he had spoken well and considered himself successful: at Scouts, at a school board meeting, at town council, and at the Little League banquet. He began to remember times he'd been successful instead of times he'd felt ill-at-ease.

"Try something with me, Mr. Jones. Imagine the room and the people at your meeting. Can you see them? Good. Now, listen to your voice. Hear how strong it is? Listen to the others. You know them well, and they *always* answer when you talk to them, just as they are answering in the meeting. See the outline of the meeting on the paper in front of you? Listen to how enthusiastic they are! They say they knew you'd be a good leader, and they're looking forward to next week's meeting."

"Hey, yeah, I see it! I could do okay, couldn't I? That's just like with the Scouts or something; if you have something to say and don't act biggety, people will listen, then talk with you about it. Hey, I really can do this thing!"

As the personnel manager, his peers, and his wife knew he would, Mr. Jones became an outstanding quality-circle leader. By the time he met with management, he was so convinced of the worth of his group's ideas that he wasn't bothered at all about the correctness of his syntax.

Analysis

By helping Mr. Jones learn positive self-talk, the manager was able to

1. Encourage him to express his what-ifs and turn them into positive instead of negative thoughts.
2. Use past success to help Mr. Jones change his thoughts and feelings to envision action leading to future success.
3. Help Mr. Jones image himself as a leader through a positive scenario of success.
4. Change negative thoughts to positive thoughts. Encouraging positive self-talk set Mr. Jones up for a good experience. He anticipated success and therefore was confident enough to be successful.
5. Reinforce thoughts and feelings with action. The manager not only taught Mr. Jones positive self-talk but also provided him with the skills he needed to be an effective group leader through the quality-circle leadership training—an unbeatable combination!

Practice

1. Find a person who agrees to help you practice working with a client on positive self-talk. Assume that you and your coached client have previously conceptualized the client's concerns.

2. Select one situation from the list of role-plays that follows these instructions. Before starting your mini-interviews, clarify specifics of the situation: age and gender of the client, setting of the interview, and other aspects essential to getting started. Work with your client and decide on details related to *negative* self-talk in a particular situation.

3. Conduct a ten-minute mini-interview for each role-playing situation. Record your interaction or have another person observe the interview, especially noting aspects related to the helper's response to positive and negative self-talk. To start the interview, either the client or you should summarize the client's concern.

4. Following each mini-interview, ask your coached client for feedback about the interaction. Then listen to the tape or talk to your observer and answer the questions in the "Focus Checklist," which follows the role-playing situations.

5. Try some mini-interviews in which you practice appropriately integrating the techniques you learned in Parts 1–3 with use of positive self-talk. You can create your own role-playing situations or use any of the situations presented in the first two parts of this book. Use the "Summary Checklist" and record all appropriate data from these more integrated interviews. The "Summary Checklist" and its instructions follow the "Focus Checklist."

Role play

1. You are a personnel manager in a service-oriented business where one of the salespersons is concerned with a slump in personal sales record. The company is not concerned because such declines happen to all salespersons, but it is evident that the young sales representative is losing confidence. The director of marketing has asked you to help rebuild this young person's self-confidence, hopefully in time to meet with an important customer who is due into town soon.

2. You are a hospice volunteer meeting with a widow who sees no hope for future social contacts now that she is no longer half of a couple. You must help her recognize present and future possibilities and restore her confidence in her own ability to meet and interact on her own with strangers and current acquaintances.

Focus checklist

1. How did you introduce the idea of positive and negative self-talk to the client?
2. Did you assist the client to identify negative self-talk, the *what-ifs,* during the interview? Did you catch all of the negatives, both stated and implied?
3. Does writing down specifics help pinpoint aspects of the negative self-talk?
4. Did the client, with your aid, turn negative thoughts into positive ones?

5. How did you assist the client to build a positive image of future success?
6. Did you assist the client to envision realistic success, or did the future become so rosy that failure seemed likely?

Summary checklist

How many of the following elements did you use, and in what ways did you use them during your more integrated mini-interviews?

1. Structuring the helping relationship
2. Focus on thoughts, feelings, actions, or TFA combinations
3. Listening for understanding and nonverbal aspects
4. Asking open questions
5. Asking closed questions
6. Systematic inquiry
7. Clarification
8. Reflection
9. Use of silence
10. Confrontation
11. Client concerns
12. Goals
13. Procedures
14. Evaluation
15. Termination and follow-up

Instructions

To assess your interview skills, make a worksheet similar to the Behavior Summary Checklist. In the *left column* on the checklist, make a transcript of key words and phrases or sentences used by you and your client during the mini-interview. Then, *for each* transcript response, put a check in columns that are relevant. You might check from one to several items for each response.

When you have completed the Behavior Summary Checklist, review it carefully. Look for patterns where you emphasized or ignored content. Then write a brief summary of the mini-interview outlining your strengths, limitations, and suggestions for improving skills.

References and recommended reading

Brouwers, M. (1990). Treatment of body image dissatisfaction among women with bulimia nervosa. *Journal of Counseling and Development, 69*(2), 144–148. This research speaks to changing women's perceptions of body image, associated with eating disorders. Treatment of eating disorders requires training beyond the scope of this text.

Summary Behavior Checklist

Key words and phrases of the Client (CL) and the Helper (H)

Focus or Emphasis: T	
Focus or Emphasis: F	
Focus or Emphasis: A	
Structure	
Nonverbal aspect	
Open question	
Closed question	
Systematic inquiry	
Clarification	
Reflection	
Confrontation	
Use of silence	
Client's concern	
Goals	
Procedures	
Evaluation	
Termination	
Followup	
Strategies	

Ellis, A. (1973). Rational-emotive therapy. In R. Corsini (Ed.), *Current psychotherapies*. Itasca, IL: Peacock. Ellis's theory of rational-emotive therapy is one of the basic theories of counseling. Much training and supervision are needed to become a rational-emotive therapist.

Gilliland, B. E., James, R. K., Roberts, G. T., & Bowman, J. T. (1984). *Theories and strategies in counseling and psychotherapy*. Englewood Cliffs, NJ: Prentice-Hall. Chapter 9, "Reality Therapy," includes a section on being positive, with a discussion of breaking negative patterns and replacing them with positive behaviors.

Hackney, H., & Cormier, L. S. (1979). *Counseling strategies and objectives* (2nd ed.). Englewood Cliffs, NJ: Prentice-Hall. One of the strategies for problem intervention, discussed very well in this book, is thought-stopping.

Maultsby, M. C., Jr. (1984). *Rational behavior therapy*. Englewood Cliffs, NJ: Prentice-Hall. This book contains a complete discussion of faulty thinking and its effect on feelings and actions, with directions for therapy.

———. (1987). *A million dollars for your hangover*. Lexington, KY: Rational Self-Help Books. Maultsby discusses reforming faulty thoughts that cause faulty emotional responses.

Meier, D. (1984). Imagine that. *Training and Development Journal*, *38*(5), 26–29. This article is a discussion of many uses of mental imagery.

Rockett, G., & McMinn, K. (1990). You can never be too rich or too thin: How advertising influences body image. *Journal of College Student Development*, *31*(3), 278. These authors describe a program used by a university to influence how students develop body image through advertising influences.

Schaub, B. G., & Schaub, R. (1990). The use of mental imagery techniques in psychodynamic psychotherapy. *Journal of Mental Health Counseling*, *12*(4), 405–415. The techniques described extend beyond those used by a beginning helper but may be of interest as helpers plan further training (Chapter 23).

Vollmer, F. (1984). Expectancy and academic achievement. *Motivation and Emotion*, *8*(1), 67–76. The author discusses the relationship among high achievement, high expectancy, and self-confidence in a university setting.

Webb, J. T., Meckstroth, E. A., & Tolan, S. S. (1983). Stress management: Some specific suggestions. *The Creative Child and Adult Quarterly*, *8*(4), 216–220. As a part of making decisions, perceiving choices, and reducing stress, the authors advocate active ignoring, which they describe as follows: "by actively filling your awareness with one thought, you are actively ignoring another" (p. 219).

Imitation and Modeling

20

Key points

1. A client can observe successful behavior in others and imitate that behavior.
2. By observing others, a client can learn how to use in new settings behaviors already learned.
3. A client can learn new behaviors by imitating others.
4. A helper can model behavior a client needs to learn.
5. A helper is often a behavior model, consciously or unconsciously, for colleagues in a work setting.

PERHAPS the most natural way for learning to occur is by imitating someone else. Just as a young child learns to tie shoes by watching an older sibling, so can a teenager learn to imitate a conversation by observing how a more successful peer interacts with others. Any social setting is a laboratory of human behavior in which a client can observe both successful and unsuccessful behavior.

Uses of imitation and modeling

A helper can use simulated or real-life examples to teach a client new ways of behaving.

1. A helper can assist a client in identifying and becoming aware of effective and successful behavior in others. The client may already possess the skills necessary for success but may never have applied them in a particular setting. Noticing how someone else is effective can teach a client a new way of using already-owned skills.

- CLIENT: I've been watching the other engineers to see how they talk to the drafters. Seems like they give fewer instructions than I do but get better results. Do you think it's possible to give too much direction to people, even if you're their supervisor?

Thus, by observing how a peer is successful, a client can see how to change personal behavior to become more effective. The *model* (the one being imitated) may or may not be aware of helping another to learn. A client may just closely watch someone who is considered effective and is therefore a subject for imitation.

One extremely successful executive relates that he carefully observes people one or two steps up the ladder to see how effective people function; practices looking at the situation, people, and processes; and rehearses potential ways of responding to the variety of situations *as if* he were in that position. When the next opportunity occurs, he has already had *experience*—mental practice—at this new level of responsibility. Or the client or helper may talk with the model to enlist the help of the person.

- HELPER: You say that Jane is the best student in the class and that you're the worst. I want you to watch what Jane does in there for the next few days. When does she watch the teacher? Does she ever let the teacher know she needs help? What does she bring with her to class? How can you tell she's paying attention? Make a list of what you observe that seems to work well for her. If you feel comfortable doing so, ask her how she studies for tests, what she does for homework each night, and how much time she spends each day on the subject. Next week bring in your list and you can decide if you want to try any of those things.

2. The helper can take the role of model. Showing how to maintain eye contact, give a firm handshake, or use the process of decision making can teach a client just how something can be done. The helper must be aware of being observed by the client—and by colleagues—in the work setting at all times. Exploding in anger at a meeting may be very human but does not demonstrate rational problem solving for colleagues!

Demonstrating the difference between various ways of saying or doing something can help the client see differences that might not be clear by just talking about them.

- CLIENT: Now I see what you mean about being assertive. Until I heard you say the same thing in two different ways, I didn't really understand the difference between being my usual mealy-mouthed self and being a more assertive person.

3. Imitation can be useful to someone with a disabling condition and may become an effective coping strategy. A person with a physical, emotional, or learning problem may be able to go through the motions when following another person but may be intellectually or physically unable to learn or carry through new behavior otherwise.

- CLIENT: I never thought, with my hearing loss, I could be a team athlete because I can't hear well enough to follow directions or understand what my teammates say. But I get up on the blocks at a swim meet, follow commands a split second after I see what the others do, and do just great—I just have to swim a little faster. And it's great just to be on the team.

Limitations of imitation and modeling

1. The client can learn by imitating only if a suitable model exists. The helper must help determine whether behaviors that exist in a model are indeed appropriate for a particular client. Enabling a shy client to turn into an aggressive bully will not help overall social interaction.

2. The behavior must be practiced so that there are not negative consequences as a result of the client's awkward application of a newly learned behavior. A client who understands the importance of eye contact in a social

setting must also understand that too much eye contact may be viewed as threatening or an unwelcome attempt to establish intimacy.

3. As with all helping situations, the client must be willing to try new behaviors. A client unwilling to risk trying new behaviors will find little value in merely observing a model. Role playing or behavior rehearsal (Chapter 21) will probably have to occur in the helper's office for some time before the client moves to a social setting. The client must *observe* and then *practice* the new behavior if it is to be used in the client's normal setting.

4. A model may be so overwhelmingly successful that the client feels unable to attempt imitation. The helper's role is, in part, to help the client distinguish explicit behaviors and break the goal into small, achievable tasks or procedures—subgoals—and then to help the client note progress toward the goal after achievement of the smaller tasks (see Chapter 12 on procedures).

5. Some clients may object to imitation. They may want to be the "real thing" and see imitating another person as trying to be someone else. The helper must help such a client understand that the process does not involve assuming another's personality but rather learning some of that person's (the model's) successful behaviors.

Imitation, modeling, and the change cycle

Goal setting may involve identifying successful behaviors in a model. The greatest use of these strategies, however, is in the procedure phase of the behavior cycle. The client and helper have clearly identified the problem and goal—what needs to be done. Imitation and modeling are two strategies that can be extremely effective in helping the client to (1) become *aware* of successful (and ineffective) behaviors and (2) learn *how* to engage in the desired behavior. When coupled with practice—both in and outside the helping interview—there is probably no more effective way of learning to *do* a new behavior.

Case study: Using imitation and modeling

As the young man shuffled into the University Counseling Center, he mumbled, "Do you, uh, treat staff here?" Receiving an affirmative answer, he continued, "Well, uh, actually, I'm an assistant professor and I just, uh, thought maybe I needed. . ."; his voice trailed away, and the man appeared poised for departure.

"Wait!" said the receptionist. "Dr. Freeman has had a cancellation and can probably see you right now." Down the hall, the receptionist said to the helper, "Dr. Freeman, I think you should see this professor out there right away before he fades off into the ivy. He's so shy he won't even look at me, and I can hardly even hear what he says when he talks."

Dr. Freeman saw the hunched figure of the young man. Extending a hand, the helper noted the downcast glance and the limp handshake from the client who identified himself as Dr. Poole. After introductory remarks, the client said, "This is my first university position, and I'm not sure I'm going to, uh, make it."

"Make it?" echoed Dr. Freeman.

"Yeah, uh, I just don't seem to have the hang of getting to know people. I'd like to be able to kind of, you know, be one of the guys—and hopefully get to know some girls, too."

"Tell me more about how you'd like to change," said the helper.

"Well, uh, I do okay at academic stuff. It's no problem for me to give a lecture because I really love my area, the Victorian Age. Sometimes I feel, uh, you'll think this is silly . . ."

"No feelings are silly in here. Go ahead," prompted the helper.

"Well, uh, it was so formal then and everyone sort of behaved by rules in public, and I, uh, think I could have done better then. Now everything is so informal, and women are so aggressive. If I can't make a friend my own age or ask a woman for a date, how can I ever get along with the department head?"

"Do you have a mental picture of how you'd like to be as a college professor?" asked the helper.

"Oh yes, I'd like to be outgoing and friendly and make small talk and be the kind of fellow everyone invites over to dinner. Or maybe ask a colleague—man or woman—out for a glass of wine after work. There's another assistant professor who seems to do all of these things, and I'm afraid he'll get tenure and I won't even be able to compete with him. Besides, I'd like to, uh, maybe be his friend."

"Can you look at me?" asked the helper. "That's *very* good! You seem much more sure of yourself when you look straight at me. Besides, you have bright blue eyes, and they add intelligence to your face."

"Oh, my, really?" the client blushed and fidgeted. "Yes, uh, I can do that, sometimes, anyway."

"Great," said the helper. "I want you to practice looking directly at others this week. This is homework, you see. I also want you to observe the other man you mentioned. Do it unobtrusively, of course. Watch how he talks to people, listen to what he says and how he says it. See how close or far away he stands. Watch how he uses his eyes, when he shakes hands, how he takes someone by the arm, whatever. Can you do that, and jot down some notes when you get home?"

"Right, yes, I can do that. And I'll, uh, try to look at people when I talk to them. Will this work, Dr. Freeman?"

"It will if you're willing to practice and learn new ways of behaving," said the helper. "Just as you learned to write a term paper step by step, so can you learn to engage in social conversation. You can imitate the behavior of others, including me, and practice here in the office. And changing how you act will change how you feel and think about yourself. Chances are, that will give you enough confidence to try establishing some real friendships. After you get some basic skills, you can work on better relationships with women."

Analysis

In working with a very shy client who could learn by imitating someone else, the helper did these things:

1. As with any hesitant client, the helper saw the client as soon as possible—before he could change his mind.
2. Before suggesting the technique of imitating, the helper first took the client through the basic helping process of clarifying concerns and setting goals. The client could describe *desirable* behavior. Then the helper noted the client's identification of a model.
3. The helper *modeled* typical social behavior by maintaining eye contact, shaking hands, and making introductory small talk. The helper pointed out one facet of this social behavior, eye contact, that would continue to be a model for the client at later sessions.
4. The client received a *homework* assignment and agreed to complete it: practicing eye contact and noting another's behavior.
5. Observation of the model was structured for the client. The helper *detailed just what behaviors to look for* and even suggested note taking, a comfortable method of gaining new knowledge for the professor.
6. The helper related the *single task* of establishing eye contact to the larger goal of maintaining social relationships and explained to the client the interrelationship among thinking, feeling, and acting.

Practice

1. Find a person who agrees to help you learn how to use the strategy of imitation and modeling. Assume that you and your coached client have previously conceptualized the client's concerns.

2. Select one situation from the list of role-plays that follows these instructions. Before starting your mini-interviews, clarify specifics of the situation: age and gender of the client, setting of the interview, and other aspects essential to getting started.

3. Conduct a ten-minute mini-interview for each role-playing situation. Record your interaction or have another person observe the interview, especially noting aspects related to the helper's response to use of modeling and imitation. To start the interview, either the client or you should summarize the client's concern.

4. Following each mini-interview, ask your coached client for feedback about the interaction. Then listen to the tape or talk to your observer and answer the questions in the "Focus Checklist," which follows the role-playing situations.

5. Try some mini-interviews in which you practice appropriately integrating the techniques you learned in Parts 1–3 with use of modeling. You can create your own role-playing situations or use any of the situations presented in the first two parts of this book. Use the "Summary Checklist" and record all appropriate data from these more integrated interviews. The "Summary Checklist" and its instructions follow the "Focus Checklist."

Role play

1. Your client is interested in learning to meet and interact more effectively with unknown persons in a social situation. Help this person develop a plan for using appropriate eye contact and physical expressions for greeting people. Model what to do to initiate and maintain a good conversation.

2. You are working with a person who is determined to give up smoking and has identified a co-worker who has successfully quit smoking. Explore with your client the possibility of using that co-worker as a model. How would your client approach the model for help?

3. A parent has noticed that a neighbor seems much more effective in handling children's anxieties and fears. You suggest that your client may learn by watching the neighbor and by asking the neighbor how to handle various situations. Structure the observation and learning process for your client.

4. Your client is interested in becoming a secretary/receptionist at a major company. Work through how the client can observe the current successful person's behavior to learn what behavior needs to be developed. Make a list of these items.

Focus checklist

1. How did you explain to the client the purpose of the technique and how it fitted in with an overall helping process?
2. Did you give clear and complete instructions on how the client was to observe and then imitate?
3. How did you personally model behavior?
4. What kind of feedback about efforts at imitating the model did you give the client?
5. How did you relate modeling and imitation to the client's real-life situation? Did you give instruction about practicing new behavior?

Summary checklist

How many of the following elements did you use, and in what ways did you use them during your more integrated mini-interviews?

1. Structuring the helping relationship
2. Focus on thoughts, feelings, actions, or TFA combinations
3. Listening for understanding and nonverbal aspects
4. Asking open questions
5. Asking closed questions
6. Systematic inquiry
7. Clarification
8. Reflection
9. Use of silence
10. Confrontation

11. Client concerns
12. Goals
13. Procedures
14. Evaluation
15. Termination and follow-up

Instructions

To assess your interview skills, make a worksheet similar to the Behavior Summary Checklist. In the *left column* on the checklist, make a transcript of key words and phrases or sentences used by you and your client during the mini-interview. Then, *for each* transcript response, put a check in columns that are relevant. You might check from one to several items for each response.

When you have completed the Behavior Summary Checklist, review it carefully. Look for patterns where you emphasized or ignored content. Then write a brief summary of the mini-interview outlining your strengths, limitations, and suggestions for improving skills.

References and recommended reading

Crosbie-Burnett, M., & Pulvino, C. J. (1990). Pro-tech: A multimodal group intervention for children with reluctance to use computers. *Elementary School Guidance and Counseling, 24*(4), 272–280. The authors used older children as "pro-tech models" for teaching younger children how to use computers.

Cull, J. G., & Golden, L. B. (1984). *Psychotherapeutic techniques in school psychology*. Springfield, IL: Thomas. The authors discuss modeling as a part of a peer-counseling program.

Gumaer, J. (1984). *Counseling and therapy for children*. New York: Free Press. The author describes modeling behaviors as a counseling tool, as well as behavior rehearsal.

Hackney, H., & Cormier, L. S. (1988). *Counseling strategies and interventions* (3rd ed.). Englewood Cliffs, NJ: Prentice-Hall. Chapter Eleven, "Using Counseling Interventions," contains a section on social modeling.

Rickel, A. U., Gerrard, M., & Iscoe, I. (1984). *Social and psychological problems of women: Prevention and crisis intervention*. Washington, DC: Hemisphere. This book includes a discussion of role models and their importance in women's development.

Summary Behavior Checklist

Key words and phrases of the Client (CL) and the Helper (H)

Focus or Emphasis — T	
Focus or Emphasis — F	
Focus or Emphasis — A	
Structure	
Nonverbal aspect	
Open question	
Closed question	
Systematic inquiry	
Clarification	
Reflection	
Confrontation	
Use of silence	
Client's concern	
Goals	
Procedures	
Evaluation	
Termination	
Followup	
Strategies	

● Role Playing

21

Key points

1. Role playing—trying out a new behavior—lets a client experience a new way of thinking, feeling, and acting.
2. Role reversal occurs when two or more persons exchange roles and each tries out the other's behaviors.
3. Behavior rehearsal lets a client practice a new behavior in the safety of the helper's office.
4. Role playing often gives clients a new perspective of their own behavior, as well as allowing them to see others in a different way.
5. Videotaping or audiotaping can show a client a new view of self during a role play, as can feedback from the helper and from other members of a helping group.
6. Role playing allows a client to try out new behaviors observed in a model.

CHILDREN try on different personalities easily as they play at being nurses, mommies and daddies, professional football players, cowboys, and various other adult roles. As we grow up, we lose the capacity to step naturally into another role and see how it feels. Helpers can assist clients to gain perspective through other eyes by playing roles dissimilar to those they fill during normal daily lives.

To play a role is to assume a viewpoint or perform actions different from one's usual routine. Becoming a different person for a few minutes has many benefits. It can help a client experience unfamiliar thoughts, feelings, or actions though simulation; understand a different viewpoint; practice a new behavior observed in a model (Chapter 20); or gain perspective about one's own present behavior.

Types of role playing

The simplest type of role playing is assuming a different role than usual—a welfare client practicing a job interview, an eighth grader playing a part in a career simulation, a college freshman participating in an orientation program. The client pretends to do something not usually done: engaging in different actions and feeling what it is like to be another person or to be in another setting for a few minutes.

- CLIENT: I'm so glad we role-played the interview before I went to the personnel manager. I'd forgotten all those questions they ask, and it's been twenty years since I came up with some of those answers. I'll be much less nervous than I would have been if I hadn't had some idea of what to expect as a result of our practice session.

Role reversal

Role reversal may involve one person (client) with the helper, with an empty chair, or with another member of a helping group. The client reverses the role usually played, maybe moving to the empty chair to symbolize being another person for a few minutes. Two or more persons can also exchange places. If there are two or more clients exchanging roles, actually shifting into each other's chairs will enhance each one's feeling of having assumed the other's role. Sometimes a hat or a sign is useful to distinguish which role is which. A client may assume a role opposite to the usual role, often of a person with whom the client is in conflict. Parent and child, teacher and student, supervisor and supervisee may change sides of an issue, each trying on the other's perspective for a few minutes in the psychologically safe setting of the helper's office. A husband and wife in marriage counseling may exchange sides of an issue, each playing the other's part in an attempt to experience each other's thoughts and feelings.

- HELPER: Harry, I want you to pretend to be Alice for a few minutes. I want you to say what you would say if you were really Alice. For about the next ten minutes, you *are* Alice. And Alice, I want you to act as if you were Harry.

Role playing in the helping group

The strategy of role playing is particularly useful in helping groups. The group can support the client who is trying out new behavior through role play. Two members in the group can switch roles, or two can role-play opposite each other—for example, one playing a parent and one a child. The goal is to have the client walk in another's shoes, to experience a different, often conflicting, point of view. The client can often understand another perception of a situation and may modify views to gain insight and perhaps reduce conflict.

- CLIENT: I guess I never really thought about what it was like not to have any friends and to talk funny like he does. I still don't like him, but I'll quit picking at him and I'll tell the others to lay off, too. I didn't like it when I was pretending to be him. I'd much rather be me.

Behavior rehearsal

Behavior rehearsal involves a client trying out a new set of behaviors that the client would like to perform. The helping setting provides a safe environment for practicing new behavior—trying it out to see how it feels and actually practicing and reviewing it until it becomes comfortable or at least achievable. Sometimes the practice is videotaped so that the client can view the new role; for some clients, however, this is too threatening, at least initially. A client who wants to learn to make friends, for example, can practice making eye contact, initiating a conversation, inviting a friend to do something, adding or eliminating mannerisms, raising or lowering the voice, and other such social behaviors necessary for contact with another person. Another client may practice making assertive statements and avoiding submissive statements and movements. The helper can observe the new behavior and offer feedback, giving the client a chance to attempt and practice new behaviors before getting into a real-life situation where correction by peers might have more damaging effects. It is more productive to practice assertive statements with a skilled and understanding helper who can give both positive and negative feedback than with a bullying spouse who habitually overrides objections. A supportive helping group is a good place to rehearse new behavior, receiving feedback from both the helper and group members.

Using videotape or audiotape is an effective way of monitoring role rehearsal. Especially with videotape, the client can see and hear the new behavior and can critique performance. In a group setting, other members of the group can also offer useful feedback to a person learning a new behavior. The client can take the videotape home and study it at leisure, getting used to seeing self in a new role and continuing to image new behaviors (Chapter 20).

- HELPER: You sounded so much more definite that time when you spoke. You almost shouted at me: I sure felt you were determined to get an answer and were not about to back off.

Five steps for role playing

Setting

Set up the situation clearly. The setting should be focused on the particular situation the client has experienced or will be experiencing. Who is there? What is the physical setting (in a hallway, at dinner, at work in a crisis situation, in a locker room, etc.)?

Roles

Assign the roles. Make certain the client understands the role to play: self, other person, reversal of normal role, and so on. Emphasize that feelings and actions should be used, thereby providing maximum understanding of emotions involved in the role.

Time

Role-play only as long as needed. Sometimes only three or four minutes are needed to demonstrate the major points or to illustrate concepts to the client. The helper should stop the role play when the points have been made, rather than letting it drag on.

Observation

Elicit observations from the client. What were the client's thoughts, feelings, and actions in the role-play situation? If there is a videotape, this is a good time to replay it.

Implications

Discuss implications of the role play. Did it result in a greater cognitive or emotional understanding or demonstration of actions that need to be taken? If so, what are these? How does the client need to change? Can the client practice behavior to more skillfully use techniques in real life? Which skills? How? Where? When?

Uses of role playing

Role playing gives people a perspective on their own behavior. If, for example, a child plays the teacher while the helper plays the child, the youngster may understand much better how student behavior causes certain reactions in the teacher and among classmates. Playing the role of the teacher may help the child sense the adult's frustration in dealing with students who are disruptive or disrespectful.

Through role playing, a client rehearsing a new behavior will be able to feel some of the same reactions that will be present when the behavior is tried out for real. If it is embarrassing to look directly at the helper or at group members, then establishing eye contact in a real job interview, outside the low-risk setting of the helper's office, will probably be even more embarrassing and uncomfortable. The helper and client can consider the feelings associated with the action so that the client's new behavior has a better chance of success when used in an actual setting. Also, practice enhances confidence in the behavior; the new behavior feels more comfortable after practice and thus becomes more natural-looking and more convincing.

- CLIENT: Okay, I guess if I can do it in the group here, I can go up to somebody outside the group and start a conversation and maybe eventually even ask for a date. I'll try, anyway, and if it doesn't work, I'll practice some more on all of you here in the group next time. I'll practice starting conversations by looking at myself in the mirror every morning and asking some questions that I've rehearsed.

Role plays can also help people understand and practice behaviors of another culture. A client may, for example, come from a culture where direct eye contact is interpreted as a challenge to authority or as disrespectful. Practicing eye contact in a group can help the client feel more comfortable in a culture where eye contact implies forthrightness, trust, and assertiveness.

Limitations of role playing

Role playing has some limitations to its applications, although it is probably one of the most useful and versatile techniques in the helper's repertoire.

1. When role playing, a person projects self into another's place, or places self into another setting. Some persons seem unable to do this. Those who will not relinquish control of themselves enough to project into the future or those who lack sufficient mental ability to conceptualize another role may be poor candidates for role playing.
2. Clear explanation by the helper of what is expected of the role player

is essential. Suppose that a helper asks a student to assume a teacher's role. If the helper fails to clearly explain how to pretend to be another person, the student may have to be reminded several times to stay in the role. Even with the best explanation, the student will likely assume the role and then slip out of it after a few minutes; at that time, the helper should stop the role play and once again explain what the student is to do in assuming the teacher's role. The client must really think, feel, and act the role in order for learning or changed perception to occur.

3. Some individuals are willing to take the risk of trying to learn new behavior. For others, however, trying out a new thought, feeling, or action, especially in a group setting, may be too uncomfortable. The helper must make the setting for role playing as psychologically secure as possible for the client. In such cases, individual practice is desirable before practice in a group setting. Providing this degree of psychological safety and reassurance may be beyond the skill of the inexperienced helper. For some individuals, the necessary degree of support cannot be readily provided. These clients are not initially good candidates for role playing.

4. Good feedback given to a person who has played a role is critical. Feedback (Chapter 3)—reaction to what has just happened—must be clear, specific, and encouraging and usually should contain suggestions for improvement. If the helper gives incorrect, unhelpful, insufficient, or confounding feedback, the role player may not have a clear perception of self in the new role. Both the role player and the feedback-givers need clear instruction about how they should proceed with the exercise.

5. If behavior change is in order, the helper must be sure that the client can perform with some success in the role assigned. If the client cannot do what is asked, further discouragement and a lessening of the current level of skill might result. A client should not be asked to play a role beyond personal capability.

Role playing and the change cycle

Role playing is probably one of the most useful strategies in the helper's repertoire. In the assessment phase, role play can be used to evaluate a client's level of experience, success, skill level, and comfort with a particular behavior after the helper observes the client in role play. There is almost no better way of assessing *how* the client engages in various kinds of interaction than to role-play the situation.

In the procedures phase, role play can be used to change thoughts, feelings, or actions by giving the client a way to try out and practice new behaviors within the safety of the helping setting. As a result of role play, the helper can probe for more insight as the client tries on another viewpoint or experiences another's way of thinking, feeling, or acting, especially through role reversal. The helper can determine the suitability of a new behavior as a client tries it out in a psychologically safe setting where feedback is available and immediate corrections and suggestions can help the client improve and learn more effective behaviors quite rapidly. If the behavior is not ultimately appropriate for a client, it can be corrected and other thoughts, feelings, or actions can be substituted.

Case study: Using role playing

As Ms. Mendez faced the group of engineering managers, she was aware of their hostility. Accepting that an outside consultant was going to teach them to be better interviewers was probably difficult for some managers. However, they were distressed mostly because earlier in the month she posed as a prospective employee and had been interviewed by each of them so that she could see first-hand their strengths and weaknesses.

Praise for their interview techniques in general eased the managers' anxiety somewhat, and they openly discussed the problems they faced in trying to hire good workers. They acknowledged their need for skills and agreed to work with Ms. Mendez to learn new ways of proceeding.

"I'm going to show you a videotape of a manager interviewing a prospective engineer, and I want you to think about what you must learn to conduct an interview in a professional, efficient manner. Then I want you to split up into pairs so that each of you can role-play being the interviewer. You will practice with each other and refine your skills." The anxiety level shot up again, but the men rearranged themselves as the consultant requested. After viewing the videotape and discussing several points, Ms. Mendez gave these instructions: "Remember that we all have the same goal: to learn to interview well so prospective employees you want to hire will want to work here. No one is in competition with anyone else; your task is both to improve your own skills and to help your co-worker improve his, to benefit the company and the engineering section. This is a place to practice, to give and receive feedback openly and honestly, and to try to improve performance."

Initially, the managers were obviously uncomfortable, but as they practiced, they saw the worth of the exercises. Following Ms. Mendez's printed guidesheets, each person being interviewed made suggestions on interviewing techniques. In the general discussion afterward, comments such as the following were voiced—no surprise to Ms. Mendez who was familiar with the power of role playing as a way of improving performance in the workplace:

- "I was really impressed with how Jack seemed genuinely to care about my answers. He even asked sensitive questions, but because he seemed to care, I felt okay about answering them."
- "It's been a long time since I sat in the chair of the prospective employee. I had forgotten how tense a spot that is, even if it's just role playing. Now that I've sat in that spot again, I believe I'll be more understanding about someone's nervousness and not just think 'that guy's too uptight to work here.' I think I'll look further and try harder to put someone at ease."

- "I thought I'd feel foolish pretending to interview Sam, but the comments he made really were helpful. I can see where I need to talk less and listen more."
- "I saw myself through someone else's eyes when Ted showed me what he'd written. I wonder if we could use the videotape equipment we bought for training workers in new procedures and see ourselves role-playing on that. I'll bet that would really be helpful."

Analysis

Role playing is an excellent teaching device. It helps a client try out new actions, thoughts, and feelings. In using role playing in the industrial setting, Ms. Mendez did the following:

1. She enabled her clients to experience someone else's perspective (that of interviewee) in a way they could use to change their perceptions, if necessary, of a job applicant's feelings and actions during an interview.
2. The clients could try out new behavior in a safe setting where the only goal was to learn and improve. No prospective employees would be lost due to poor interview techniques because none were involved. The helper provided a setting for practice with peers that was neither competitive nor judgmental because all shared the same goal.
3. The helper structured the role playing so that all understood the purpose, the procedure, and the results of the exercise.
4. In this case, all participants played both roles; thus, all clients had the same experiences. The commonality of experience was designed to provide a team approach to solving the problem of better interviewing techniques.
5. The helper provided plenty of opportunity for discussing what happened when the clients played the roles. Thus, each learned from the others' experiences, multiplying the effect of individual practice.
6. The idea of videotaping was introduced, offering an even greater opportunity for each person to react to personal efforts and to improve methods of interviewing.

Practice

1. Find a person who agrees to help you learn how to use the strategy of role playing. Assume that you and your coached client have previously conceptualized the client's concerns.

2. Select one situation from the list of role-plays, which follows these instructions. Before starting your mini-interviews, clarify specifics of the situation: age and gender of the client, setting of the interview, and other aspects essential to getting started.

3. Conduct a ten-minute mini-interview for each role-playing situation. Record your interaction or have another person observe the interview, especially noting aspects related to your response to positive and negative self-talk. To start the interview, either the client or you should summarize the client's concern.

4. Following each mini-interview, ask your coached client for feedback about the interaction. Then listen to the tape or talk to your observer and answer the questions in the "Focus Checklist," which follows the role-playing situations.

5. Try some mini-interviews in which you practice appropriately integrating the techniques you learned in Parts 1–3 with use of role playing. You can create your own role-playing situations or use any of the situations presented in the first two parts of this book. Use the "Summary Checklist" and record all appropriate data from these more integrated interviews. The "Summary Checklist" and its instructions follow the "Focus Checklist."

Role play

1. You are a family counselor in a community mental health center working with a teenager who feels great conflict with a parent. Ask your client to assume the role of the parent by moving to a vacant chair. You will become the teenager while your client becomes the parent.
2. Your client is an adult trying to get a job after being out of the work force for twenty years. Role-play a job interview in which you are the interviewer and your client is applying for a job.
3. As you have done the exercises at the end of each chapter, *you* have been role-playing. Having read this chapter, examine the power of the role-playing experience for you as a learner. Do you see parallels that will apply in the future to your clients?

Focus checklist

1. How did you use role playing so that the client understood what the strategy was supposed to add to the helping process?
2. Did you give clear and complete instructions to the client?
3. Did you follow the five steps for effective role playing?
4. What kind of feedback did you give the client?
5. How did you relate role play to the client's real-life situations? What kinds of instructions did you give about use of the new behavior?

Summary checklist

How many of the following elements did you use, and in what ways did you use them during your more integrated mini-interviews?

1. Structuring the helping relationship
2. Focus on thoughts, feelings, actions, or TFA combinations
3. Listening for understanding and nonverbal aspects
4. Asking open questions
5. Asking closed questions
6. Systematic inquiry
7. Clarification
8. Reflection
9. Use of silence
10. Confrontation
11. Client concerns
12. Goals
13. Procedures
14. Evaluation
15. Termination and follow-up

Instructions

To assess your interview skills, make a worksheet similar to the Behavior Summary Checklist. In the *left column* on the checklist, make a transcript of key words and phrases or sentences used by you and your client during the mini-interview. Then, *for each* transcript response, put a check in columns that are relevant. You might check from one to several items for each response.

When you have completed the Behavior Summary Checklist, review it carefully. Look for patterns where you emphasized or ignored content. Then write a brief summary of the mini-interview outlining your strengths, limitations, and suggestions for improving skills.

References and recommended reading

Chapman, A., & Lumsdon, C. A. (1983). Outdoor development training. *Leadership and Organization Development Journal*, *4*(4), 28–31. The authors describe learning through behavior rehearsing: surveying a problem, developing and implementing solutions, and evaluating outcomes in a setting markedly different from the workplace, for transfer to the workplace.

Hackney, H., & Cormier, L. S. (1988). *Counseling strategies and objectives* (3rd ed.). Englewood Cliffs, NJ: Prentice-Hall. The authors use exercises to augment discussions of role playing, role reversal, and behavior rehearsal.

Jessop, A. L. (1979). *Nurse–patient communication: A skills approach*. North Amherst, MA: Microtraining Associates. This book offers many opportunities for role playing, to learn communication skills.

Martin, G., & Hrycaiko, D. (1983). Effective behavioral coaching: What's it all about? *Journal of Sport Psychology*, *5*(3), 8–20. The authors discuss the effect of the coach's behavior as a role model and of behavior rehearsal to improve performance.

Robinson, S. E., & Kinnier, R. T. (1985). Sex and order of counseling effects on paired

Summary Behavior Checklist

Key words and phrases of the Client (CL) and the Helper (H)

T F A

Focus or Emphasis	
T	
F	
A	
Structure	
Nonverbal aspect	
Open question	
Closed question	
Systematic inquiry	
Clarification	
Reflection	
Confrontation	
Use of silence	
Client's concern	
Goals	
Procedures	
Evaluation	
Termination	
Followup	
Strategies	

role-play practice. *Counselor Education and Supervision, 25*(2), 143–148. These researchers examine the effect of variables on role play.

Tsui, A. M., & Sammons, M. T. (1988). Group intervention with adolescent Vietnamese refugees. *Journal for Specialists in Group Work, 13*(2), 90–95. This article illustrates the use of role play by a helper in working with Vietnamese refugees.

Using Additional Strategies

22

Key points

Bibliotherapy
Nonprint materials
Computer programs
Toys and other manipulatives
Games
Play therapy
Testing

THIS chapter contains a compendium of other strategies, with a short description of each. For those strategies noted with an asterisk, the novice helper will want to look for further training through coursework, conferences, intensive workshops, reading, or other sources of additional training. This chapter does not contain the practice exercises, which we hope that you have found helpful in other chapters. References for further reading are listed with each strategy.

Bibliotherapy

The helping process usually consists of talking: the client voices thoughts, relates past and present feelings, and charts future action. The helper asks questions to gain content, clarifies and summarizes, and responds as the client plans strategies, sets goals, and assesses progress. Previous chapters have outlined the use of writing in the helping relationship. For some clients, reading can be a very useful addition to talking and writing as a part of the helping process.

Bibliotherapy is the use of the printed word to assist in the helping process. By assigning reading matter to the client, knowledge, understanding, and discussion of the written material become a part of the helping process. There are hundreds of self-help books on the market. As a helper, you may have several of these, which you can loan to clients. It is helpful to recommend *chapters* or *short sections* to clients who can benefit from specific reading to reinforce or supplement other helping strategies. For some clients who are unused to revealing themselves, reading a particular book or article can provide insights that unlock thoughts, assure them that other people have the same or similar concerns, and lay the foundation for helping to proceed verbally.

- CLIENT: This character in the book says exactly how I feel most of the time.

Other clients may use books to gain information they need to use in decision making. Although skilled in the decision-making process, no one helper has all the information various clients need to make their own decisions on different topics.

Sometimes clients can identify with characters in a work of fiction or with feelings expressed in a poem or in the lyrics of a song. Exploration of those feelings can be enhanced by the helper and client discussing the client's reaction to a main character or a line of a song.

- CLIENT: That book *Tales of a Fourth Grade Nothing* by Judy Blume is about *me!*

The following is a sampling of material that may be useful in bibliotherapy:

Blume J. (1974). *Tales of a fourth grade nothing*. New York: Dell. Blume depicts a typical fourth-grade self-concept.

Cook, J. A. (1983). A death in the family: Parental bereavement in the first year. *Suicide and Life-Threatening Behavior, 13*(4), 42–61. An extensive listing of books and articles relating to death follows this article.

Dinkmeyer, D., & Losoncy, L. E. (1981). *The encouragement book*. Englewood Cliffs, NJ: Prentice-Hall. This excellent example of a self-help book can be used in conjunction with helping sessions.

Dyer, W. W. (1977). *Pulling your own strings*. New York: Avon Books. To help people change their image of themselves, Dyer discusses locus of control and being in charge of one's life.

Gerstein, M., & Lichtman, M. (1990). *The best for our kids: Exemplary elementary guidance and counseling programs*. Alexandria, VA: AACD. An excellent resource for helpers, this book provides the best elementary guidance and counseling programs in the United States.

Getz, H. G. (1986). Family counseling. In C. H. Humes (Ed.). *Fundamentals of counseling services*. Muncie, IN: Accelerated Development. This chapter gives a good overview of options for helping from a family systems perspective.

Harris, A. (1984, August). Books to help kids cope. *Working Woman*, pp. 106–108. The article lists many books for children, primarily fiction, to help them understand their feelings about their working parents.

Holton, E. (1991). *The new professional: Everything you need to know for a great first year on the job*. Princeton, NJ: Peterson's Guides.

McDaniels, C. (1978). *Developing a professional vita or resume*. Garrett Park, MD: Garrett Park Press. Helpers in many different settings need information on career preparation.

Richards, A., & Willis, A. (1976). *How to get it together when your parents are coming apart*. New York: Bantam Books. Appealing to teenagers, this book can help a young person think through some things about divorce and perhaps provide courage to talk to a helper.

Materials about bibliotherapy include these:

Bloch, D. P. (1990). Integrating the school experience: A message from William Butler Yeats to counselors. *Journal of Humanistic Education and Development, 28*(4), 176–184. The author shows how poetry can be used in the helping setting.

Byrd, K., Williamson, W., & Byrd, D. (1986). Literary characters who are disabled. *Rehabilitation Counseling Bulletin, 30*(1), 57–61. The authors show how literature portrays disabled people, with implications for rehabilitation counselors.

Coleman, M., & Ganong, L. H. (1990). The uses of juvenile fiction and self-help books with stepfamilies. *Journal of Counseling and Development, 68*(3), 327–331. Stepfamilies can gain information about relationships from strategies listed in this article.

Timmerman, L., Martin, D., & Martin, M. (1989). Augmenting the helping relationship: The use of bibliotherapy. *The School Counselor, 36*(4), 293–297. The authors describe the use of bibliotherapy in a school setting.

Nonprint materials

In similar ways, use of audio-visual materials as sources of information, such as filmstrips, videotapes, and laser videodiscs, can be a very helpful strategy.

Filmstrips, tapes, films, key-sort kits, and laser videodiscs on topics such as job-search skills, employment trends and outlooks, occupational requirements, licensing procedures, and other such occupational information can be found in public and school libraries, state employment agencies, and some industrial settings, as well as in the Yellow Pages of the telephone book and the classified ads in the local newspaper. At the suggestion of the helper, the client can use such sources and then bring the information back for inclusion in the helping interview.

- HELPER: As you think about changing your career direction from social work to industrial personnel work, you are going to need a strong, current, complete vita. You can begin developing this using the steps outlined in this video on job-search skills, including how to write a vita.

Such information sources also provide generalized information. Some information must be given to many clients in about the same manner, even though different clients may use the information in different ways during the helping process. For example, almost all pregnant women need the same nutritional information.

- CLIENT: I've seen the videotape on nutrition during pregnancy, but I have some questions about how the drugs I take for migraine headaches fit in. Also, I travel and must eat in restaurants with clients, with alcohol usually a part of the evening. I need some help in planning how I'm going to keep myself and my baby within the nutrition guidelines that your videotape suggests.

Nonprint materials provide expertise the helper may lack. No matter how skilled a helper is in the process of examining alternatives and making decisions, no one helper can possibly be an expert in all the areas in which various clients will need help in making decisions. Requesting that a client read certain kinds of information brings an additional expert into the helping process.

- HELPER: I think your idea of sending your sons to boarding school while you two decide whether to continue your marriage is a good one. This would ease them away from some of the tension in the household. As a marriage counselor, I don't know much about private secondary schools, but we have some videodiscs that some of the schools have sent to us. You can watch them here in the guidance office, perhaps bringing your boys in with you.

References illustrating the use of media include these:

Cramer, D. W., and Baron, A., Jr. (1990). AIDS in the workplace: A workshop for university personnel. *Journal of College Student Development*, *31*(1), 80–81. The article illustrates use of a videotape in a helping setting.

Ivey, A. E. (1983). *Intentional interviewing and counseling*. Pacific Grove, CA: Brooks/Cole. Ivey discusses the use of videotape for the learning helper.

Computer programs

Many computer programs now on the market can assist the professional helper. Many national conferences now include computer fairs, which display hardware for helping professionals, and newsletters and journals have articles on useful computer techniques and software.

Computers can be used to give direct instruction to clients, such as skills in how to study. They can also be invaluable to helpers as they sort data, word-process letters, retrieve information, and do countless other tasks better than unassisted helpers.

Some of the programs available to helpers include

- Testing. Many standardized tests are now available on computer disk.
- Preparing letters of recommendation. These letters provide fill-in phrases, which can be completed with specific descriptors.
- Searching for career and educational information. These programs supply far more specific and accurate information than a helper can provide.
- Providing locally developed information for clients. Such information includes how to register for classes or how to fill out a job application.
- Many other uses. Computer applications are limited only by the helper's ingenuity, awareness of program availability, and purchasing power.

Articles on how computers are used by helpers include these:

Garis, J. W., & Niles, S. G. (1990). The separate and combined effects of SIGI or DISCOVER and a career planning course on undecided university students. *Career Development Quarterly*, *38*(3), 261–274. The authors discuss two well-known computer career-search methods.

Special Issue: Implementing computer technology in the rehabilitation process (1985). *Journal of the American Rehabilitation Counseling Association*, *28*(4).

Toys and other manipulatives

Children and some adults, especially those with articulation disabilities or those who are relatively nonverbal, feel comfortable with objects in their hands while they are talking. Other clients may be able to better demonstrate and illustrate their words using devices.

During an early interview, a helper may want to hand a client a puzzle or manipulative game to serve as an icebreaker. Usually this technique will be used with children or adolescents, but it may also give insights into adult behavior. For example,

- HELPER: Can you put the six pieces of this puzzle back into the box?

Such a comment can lend some informality to the interview, and manipulating the pieces of a plastic puzzle may put the client at ease. By watching the client try to assemble the puzzle, the helper may learn something about the client's approach to problem solving, frustration level, ability to plan and to learn from past mistakes, and comfort level at solving a task in a relatively nonthreatening atmosphere. The helper can also assess risk-taking behavior as the client decides whether to attempt the solution. Thus, a few minutes with a toy or game may give the helper valuable insights into how the client approaches a problem. For the more experienced helper trained in individual testing and appraisal techniques, the process of assembling blocks is often a part of testing to assess intelligence, brain damage, and other intellectual qualities.

Games

Board games

Watching a client play a board game with others gives the helper a chance to observe some of the characteristics just described and shows how the client interacts with others. How a client reacts to others verbally, emotionally, and physically in a game of checkers or backgammon, for example, may give the helper insight into behaviors that cause the client difficulty with peers. With the helper analyzing behavior the client exhibits, the client who speaks loudly and constantly, cheats, plays out of turn, verbally harasses the other players, makes running critical commentary on others' plays, or overturns the board in anger may learn to see how similar behaviors hinder social relationships in the real world.

- CLIENT: You guys *cheat*! You *stole* my money! You bought all the good property! I had the same rotten luck as always! I'll never play with you *cheaters* again!

Helping games

Some helping programs for children or for adults are devised as games. "Careers," for example, can take an adolescent through a semblance of career planning in a playful manner. Simulations and role-playing games give adolescents or adults situations they can role-play (see Chapter 21) and then discuss.

- CLIENT: That puzzle-solving activity where everyone had to work together without talking was one of the worst things I've ever done! There we were, racing against the other teams to put our puzzle together first, and everyone was feeling frustrated and angry with each other because we couldn't figure it out. It taught me something about how I respond to competition and how impatient I get with co-workers, even if it's just a game we're playing.

Articles illustrating some aspect of gaming or helping simulation include these:

Gass, M. A. (1990). The longitudinal effects of an adventure orientation program on the retention of students. *Journal of College Student Development, 31*(1), 33–38. The author describes an outdoor adventure-oriented program as a helping tool for retention of university students.

Sigelman, C. K., & Budd, E. C. (1986). Pictures as an aid in questioning mentally retarded persons. *Rehabilitation Counseling Bulletin, 29*(3), 173–181. For those who have trouble verbalizing, pictures may be an aid to helpers.

Stiles, K., & Kottman, T. (1990). Mutual storytelling: An intervention for depressed and suicidal children. *The School Counselor, 37*(5), 337–342. The authors relate how telling stories aids in a helping setting with young children.

Play therapy*

The use of play to change behavior or to elicit content from a client is well-established in helping literature. Play therapy requires much training and is *not* a technique for a novice helper.

Using dolls to depict family situations is a standard part of play therapy with small children. The beginning helper may use dolls in a nontherapy manner as a helping tool. Dolls, especially dolls that are anatomically correct, can be used in other ways as well with children, adolescents, and even adults. A particularly sensitive topic for helpers to talk about with children is child abuse, especially sexual abuse. Yet with increasing numbers of children being found to be abused and with laws requiring that abuse be reported to legal authorities, school counselors, nurses and physicians, mental health professionals, social workers in agencies and hospitals, lawyers, and others in helping situations must be able to elicit information from both children and abusive caregivers who seek help. Often children or adults may be unable to use the words with which to describe actions or body parts involved in abuse. Use of dolls allows the helper to obtain a description of just what happened in the incident.

- WORKSHOP LEADER: We as social workers must be able to talk to children about what has happened to them. Examine these dolls carefully as they are passed around the room. Learning how to talk to children about how to use anatomically explicit dolls may end up being the most significant part of today's workshop for you.

Puppets, figures representing family members, and other toys are used in play therapy.

The use of games, toys, and other manipulatives is somewhat limited in the helping setting. Depending on their age, social development, and situation, many clients may think that games and toys are silly and childish. For these clients, using such devices would only detract from the effectiveness of the helping interview and might even seem foolish.

Some helpers are uncomfortable with play or with particular aspects of using toys. As with other techniques, if the helper sees little value in a strategy, that helper should not use it.

Games are a simulation and only that. The helper must remember that because a client reacts a particular way in a game, it cannot necessarily be assumed that that client will behave the same way in real life. The client may be a big risk taker in the board game of "Careers" and carefully remain totally safe in real-life choices.

The following are some sources of information on using toys, games, and play therapy:

Canfield, J., & Wells, H. C. (1976). *100 ways to enhance self-concept in the classroom*. Englewood Cliffs, NJ: Prentice-Hall. The authors describe numerous counseling-type group games in this volume written for teachers but useful for helpers in various settings.

Gumaer, J. (1984). *Counseling and therapy for children*. New York: Free Press. Gumaer gives an excellent discussion of the use of play therapy with children. This will be a resource for persons who want information on the topic for further training.

Kottman, T., & Warlick, J. (1990). Adlerian play therapy. *Journal of Humanistic Education and Development*, *28*(3), 125–132. This article gives an overview of play therapy.

Parker Brothers, "Careers" game.

Testing*

Testing is a highly specialized strategy that requires considerable training beyond the summary offered in this text; however, even beginning helpers need some understanding of types and uses of tests. Useful data about clients can be gathered by using standardized and individually or locally designed tests and inventories such as the following:

- *Standardized personality inventories* compare a client's responses to normative data.
- *Achievement tests* measure learning in specified areas.
- *Ability* and *aptitude tests* measure a client's potential for success in a particular area.

- *Vocational interest inventories* evaluate a client's interests in potential career fields.
- Other kinds of tests and inventories measure *personality dimensions, study skills,* and *values* and a wide variety of other characteristics.

Gathering data about a client by using standardized test instruments is one of the best researched and most discussed techniques used in the helping fields. Historically, impetus for the helping movement came largely from the measurement of traits and factors and the interpretation of that data to clients for vocational and educational planning. The following brief summary of types of tests and their use for practitioners, plus suggestions for further reading, is designed to enable the reader of this text to pursue the topic further.[1]

Personality inventories are quite varied. They may range from the California Personality Inventory (CPI) and the Myers–Briggs Type Inventory (MBTI) to more projective assessments. Projective tests such as the Thematic Apperception Test (TAT), and Rorschach Inkblot Test may be used by professionals trained to administer and interpret them. Counselors and psychologists in private practice or mental health agencies or school psychologists will likely administer and interpret such tests; other helpers, however, should be familiar with this general type of test and be able to understand findings presented in a report prepared by another professional.

- HELPER: I want to refer you to a psychologist for some specialized testing. When that is completed, you may return to me or we may recommend that you work with the psychologist instead.

Achievement tests measure learning in a specified area. These tests may be group-administered paper-and-pencil instruments, such as the Science Research Associates Assessment Battery, or an individually administered oral test, such as the Peabody Individual Achievement Tests (PIAT). Such measures will be a part of any public school record and can provide information to agency or private-practice helpers, to personnel workers in industry, and to people in the medical profession.

- HELPER: Mr. and Mrs. Sites, I have Josh's achievement scores in math, reading, and other areas on the Wide Range Achievement Test [WRAT] and the Woodcock–Johnson Achievement Test. I want to discuss with you my thoughts on further testing to determine whether Josh has a learning disability.

Ability and *aptitude tests* are used in various helping settings. Individual intelligence tests, such as Wechsler Intelligence Scale for Children, Revised (WISC-R) or the Stanford–Binet Intelligence Test, may be used where there is

[1]In this discussion, be aware that the helper usually must have specialized training to use, administer, and interpret various kinds of tests. An individual who is not adequately prepared, trained, or licensed to give certain kinds of tests can be subject to prosecution for engaging in illegal and unethical practices.

a need for assessing general intelligence of a client. Group tests, such as the Otis–Lennon Intelligence Test, may be used to screen students for gifted programs. Measures such as the General Aptitude Test Battery (GATB) or the Armed Services Vocational Aptitude Battery (ASVAB) can be given to individuals or to groups. Ability and aptitude tests are used to predict a client's success in something: academic classes, a gifted program, a special education classroom, a particular university or curriculum, or a training program offered by an industry.

- HELPER: I'm glad you came to the recruiting office today. I have the results of your ASVAB and would like to talk to you about some Navy training programs we could offer you. Your scores are similar to those of persons who experience success in a number of areas.

Vocational interest inventories and other tests such as Crites' Career Maturity Inventory (CMI) measure a client's occupational interests or stage of career maturity. Tests such as the California Occupational Preference System (COPS) and the Strong–Campbell Interest Inventory (SCII) provide the helper with information that can help a client make career plans. Such inventories can be given to groups (e.g., all ninth graders in a school) or to individuals. They may be paper-and-pencil, care-sort systems, or career-interest inventories designed for the microcomputer, such as the Virginia VIEW (Vocational Information on Education and Work). Use of these instruments is growing as public schools, colleges, and adult-counseling services devote more attention to career planning, whether their clients be middle schoolers planning educational programs, graduate students seeking jobs, or homemakers re-entering the work force.

- HELPER: I'd like you to take this occupational-interest inventory. When you've finished, we'll score it and talk about what kinds of training programs we can find that will help you enter an occupation that interests you.

Other kinds of test or inventories that may be helpful to a client include

- A *values inventory* to help sort out career decisions.
- A *study-habits checklist* for a student failing in school.
- *Open-ended sentences* to discern client feelings.
- A test of *logical reasoning* to determine if a youngster is capable of learning at the formal level.
- A test of *visual perception* or *distractibility* to learn how a student functions at classroom tasks.
- A test of *self-esteem* to determine how a client feels about self.
- A test of *relationships* among thoughts, feelings, and actions, such as the Hutchins Behavior Inventory (HBI), to help a client gain new insights into the complex interweavings of behavior in a wide variety of specific situations.

Scores of such measures can assist the helper and client to plan for personal growth. Depending on the work setting, some helpers will give many different kinds of tests. Others will need to know how to read reports prepared by professionals and when to refer their clients to a psychometrist or other mental health worker for tests.

All helpers, however, will need some familiarity with types and uses of tests for their clients.

- HELPER: I asked you to meet together in a group for two reasons: You all said you want to make better grades, and according to the study-skills checklist you completed in English class, you seem to have similar problems with studying. I think we can work to help you improve those and some other study skills over the next few weeks.

The following are some references and recommended readings on testing:

Herr, E. L., & Cramer, S. H. (1979). *Career guidance through the life span*. Boston: Little, Brown. The chapter "Assessment in Career Guidance and Counseling" contains a discussion of individual appraisal for a very specific purpose. An excellent overview of testing is followed by descriptions of various kinds of measures useful in career counseling and is applicable in a mental health, medical, school, or rehabilitation work setting.

Hohenshil, T. H. (Guest Ed.). (1989). Counseling persons with disabilities: A ten-year update. *Journal of Counseling and Development, 68*. This entire issue is devoted to updating a wide variety of issues, counseling procedures, assessment, and implications for working with disabled persons.

Humes, C. W., Szymanski, E., & Hohenshil, T. H. (1989). Roles of counseling in enabling persons with disabilities. *Journal of Counseling and Development, 68*, 138–139. This article particularly addresses different roles involved in a helping relationship with disabled persons.

Miles, J. H. (1984). Serving the Career Guidance Needs of the Economically Disadvantaged. In N. Gysbers, *Designing Careers: Counseling to Enhance Education, Work and Leisure*. San Francisco: Jossey-Bass Publishers, 384–402.

Special Issue: Computer applications in testing and assessment (1986). *Measurement and Evaluation in Counseling and Development, 19*(1). This journal is published four times a year by the Association for Measurement and Evaluation in Counseling and Development. It reviews tests and reports on various aspects of testing related to counseling.

Pirie, P. L., Luepker, R. V., Jacobs, D. R., Brown, J. W., & Hall, N. (1983). Development and validation of self-scoring test for coronary heart disease risk. *Journal of Community Health, 9*(1), 65–79. The authors describe the development and validation of their self-scoring test, concluding that it can be a valid tool of health education.

Zunker, V. G. (1982). *Using assessment results in career counseling*. Pacific Grove, CA: Brooks/Cole. The authors review various categories of tests and their use in career counseling.

Summary

In this chapter, we have suggested a sampling of strategies that can be used to assist both client and helper in the change process. These strategies barely scratch the surface of the many methods that can be used in assessment, determination of goals, and procedures for changing behavior. As you progress in your training as a helper, we hope you will use these strategies as a springboard for continuing to learn about the many ways you can enhance the effectiveness of the helping relationship.

Where Do I Go from Here? Continuing Development for Helping Professionals

23

Key points

1. Helping professionals must continue to seek opportunities for professional development to stay abreast of new concepts and skills.
2. New knowledge for helping clients with their concerns must be integrated into the helper's existing knowledge.
3. Helpers must learn new techniques for dealing with client concerns, as well as legal implications and responsibilities.
4. Sources of new information for helping professionals include conferences, seminars, and workshops; professional literature such as journals, books, newsletters, videotapes, and computer programs; and networks with other professionals.

THIS text has been a brief introduction to key aspects involved in helping relationships and strategies associated with the behavior-change process. This chapter might be titled "The End of the Beginning" because this book is only a start. Only after a number of courses can you truly become a competent professional. We have outlined a number of content areas, issues, and suggestions to help you start thinking about your future as a helping professional.

The need for continued development

When a neophyte helping professional has studied, practiced, and mastered to at least some degree the concepts and practices included in this text, that individual may feel some degree of confidence in preparation for working with clients. Most helping professionals, however, very shortly after beginning work with clients, feel inadequately trained. Neither classroom work, however practical it may be, nor extensive internships and practicum experiences prepare the helping professional for the myriad of concerns and convolutions presented by an infinite variety of clients. The beginning helper soon learns what the experienced professional knows: Every client is different, and no one fits the textbook exactly. Thus, helping professionals must constantly update their skills and knowledge to continue to add to their repertoire of professional techniques. The best-trained, most effective helper of twenty years ago had no training in how to help anorectic adolescents or middle-aged persons with HIV indicators. Counselors are just beginning to realize the implications of youngsters in school who carry severe residual effects of their mothers' drug and/or alcohol abuse. Twenty years ago, who had the knowledge helpers use today in working with adult children of alcoholics or attention-deficit disordered youngsters?

New content and concepts

The helping professional must be alert for new information that is pertinent to working with clients. A sample of areas that have gained much attention in the past twenty or so years includes these:

Teenaged sexuality, especially attention to continuing education of young mothers, family-planning clinics in schools, abortion, and other issues. Typically, clients saw medical professionals for these concerns, if they sought information from anyone, in previous years. Now private practitioners, school counselors, and agency mental health professionals find themselves dealing with these topics on a routine basis, with younger and younger clients.

Substance abuse. The alarming rate of alcoholism and use of illegal drugs in our society means that virtually no site of helping is untouched by problems of substance abuse. From elementary school counselors to geriatric helpers, these issues require knowledge far beyond that available to helpers of an earlier era. An allied area currently receiving much attention is helping adults who were children in alcoholic homes. Also, there are increasing numbers of HIV-infected babies, mostly a result of the mother's intravenous drug abuse. A related issue of alarming proportions is the babies being born of addictive mothers. These infants, exemplified by the "crack babies," are clients of the helping profession virtually from birth, with their apparent problems of personality disorders, learning difficulties, physical disabilities, and perhaps as yet unknown other complications.

Health issues such as AIDS and other sexually transmitted diseases. Added to the pregnancy and moral issues, which helpers knew about twenty years ago, are health problems eclipsing those that required caution only a generation ago. Yet neither younger nor older clients have routinely learned the judgment and caution required to avoid these risks.

Child abuse and other forms of domestic violence. From sexual abuse of infants and children, through date rape and violence against aging relatives, the helping professional has learned that family violence is a pattern that repeatedly surfaces, regardless of the age or socioeconomic strata of the clients with whom the professional works.

Societal problems for groups of people such as Vietnam War veterans; victims of mass shootings, discrimination, or persecution; and other clients affected by life-disrupting events such as earthquake, floods, or disease that seemingly occur to random groups of clients.

Career changes that are the result of new opportunities or, conversely, unforeseen radical changes in the economy. Some theorists forecast periods of unemployment for many groups of people in the future, especially those with little education or skill to transfer to a new job in a rapidly shifting workplace. Career counseling was once regarded as the domain primarily of helpers working with adolescents or young adults. Now, virtually all helping professionals work with issues related to careers—from elementary counselors who promote career awareness and professionals who assist midlife career changes to geriatric professionals who help their clients adjust to retirement, loss of income, or dependency on society.

Another important area to the helping professional is keeping up to date on changes in the law, especially relating to aspects of the professional's practice. Most helpers know, for example, that they are required to report even suspicion of child abuse or neglect. The legal requirements may be less clear in other areas, however.

- What is the helper's legal responsibility when two clients' rights seem to collide, as may be the case in mediation proceedings?
- When does a helper warn a relative of an adult's threatened suicide?
- Who is legally entitled to see records relating to an individual, such as a child of divorced parents?
- To which government agency must a helper release information about a client?
- Who has privileged communication with a client and which helpers do not?

The answers to these and other legal questions change from time to time and vary for different locales and different work settings. The helping professional can be in deep trouble indeed if not aware of the legal requirements of a particular job and locality.

Equally important are regulations governing behavior of the practitioner in a particular work setting. Requirements for reporting a client's substance abuse, for example, may be very different for a school counselor than for a probation worker. Institutions employing helping professionals set regulations to which their employees must adhere if they want to keep their jobs. The wise beginning helper will be very aware of these regulations before accepting a job, thus avoiding the position of not being comfortable with or willing to support the regulations of the employing agency. Also worthy of consideration may be the rules or practices—written or unwritten—of a community or private agency.

- Are parents routinely informed, for example, of feelings that their child confides to a helper in a private setting? If so, does the new helper believe this to be a correct practice? Is the helper willing to comply with this expectation?
- Are substance-abuse agency counselors expected to share information, sometimes of a confidential nature, with law enforcement and court officials?
- Are ethical, moral, and religious positions of a family-planning clinic sufficiently congruent with the helper's personal beliefs so that the employment is a good choice? How does the professional react if the director changes policies and direction?

The helping professional must constantly stay tuned to trends in treatment philosophy, new areas of concern, and legal and policy requirements of the job.

Skills and techniques

Helping professionals must be aware of new developments in techniques for treatment of clients. Training changes from time to time, as researchers document the effectiveness of new methods of working with clients. Professionals receiving training twenty or thirty years ago likely had little theory or practice working with groups of clients; currently, helpers-in-training in most programs receive both theory of and practice in group counseling. One might not have predicted, for example, these techniques in counseling some years ago:

Videotaping clients as they play roles. Role playing (Chapter 21) might or might not have been taught as a technique, but surely videotape was not readily available to use to analyze behavior.

Use of *videotape as instructional material*. Replacing or augmenting bibliotherapy (Chapter 22) may be videotherapy, giving clients videotapes on various subjects to improve their information in various areas.

The multitude of uses of the *computer* in the helping professional's therapeutic office. From testing to gathering intake data to teaching skills to keeping a journal to uses not yet imagined by today's helping professional, the computer surely will have the same impact on the helping profession as a tool that it has had on other segments of the work world.

Use of *other technology*, such as the fax machine, and other new uses of communication devices, such as for transmitting information. How fast can a school get information on a new student with antisocial behavior in the classroom, or how fast can a substance-abuse counselor learn about a new client in crisis? As fast as the sending school or previous agency can transmit that information over the telephone lines.

Value of not-yet-widespread techniques such as *peer counseling*, whether with adolescents, midlife persons, or older adults. How does one set up a peer-counseling or peer support group in a community agency, for example? How do self-help groups start and sustain themselves, with a helper's support?

New theories and implications for treatment. This text is largely based on the TFA system; how does a neophyte helper learn about other new approaches to theory and technique as new ideas develop and gain acceptance?

New applications and settings for tried-and-true theories and techniques. Will helping professionals soon be routinely available in large corporations and businesses, as personnel departments change into human resource management divisions? Will nursing homes and hospitals employ large numbers of helping as well as medical personnel to work with psychological as well as medical aspects of disease, defects, disabilities, and development such as aging?

Similarly, the helping professional will find it very useful to periodically take a "refresher" course to keep skills finely honed. If one works in a setting for several years, for example, where only individual counseling is done, techniques of working with groups will be at best rusty. Some way of renewing these skills before attempting group counseling will give confidence to the helping professional and make early sessions more profitable than if all skills are regained through reuse alone.

Sources of continued professional development

Many avenues allow helping professionals to continue their growth throughout their careers. Good ways of keeping up to date are didactic and experiential presentations; journals, books, and other print material; presentations of salespeople and representatives of agencies; and networking with other professionals for mutual renewal and development.

Some helping professionals are required by their licensing board to participate in professional development activities; typically they must accrue a certain number of points or course hours within a given time period to retain their credentials. For example, in many states, a school counselor must attain a specified number of credit hours or other kinds of professional development points within five or ten years. A helping professional certified by the accrediting board of the American Association of Counseling and Development (National Board of Certified Counselors) must accrue professional development points from among many different activities (conference attendance, formal graduate courses, professional publication, and others) every five years. Similarly, a social worker credentialed by the Academy of Certified Social Workers must meet specified criteria to maintain the endorsement.

Courses, seminars, workshops, and conferences

Many helping professionals find personal and professional renewal through presentations of various types. They like not only receiving new knowledge and skills but also interacting with presenters and others attending the presentations. Among these resources are the following:

Annual conferences and conventions—local, state, regional, national, and international. Professional organizations typically offer some kind of professional development credit to their members who attend sessions at these conferences. Often, state or national conferences offer a fine combination of well-known speakers in the field, coupled with sessions delivered by effective practitioners.

Short-term seminars and workshops, often offered by an organization interested in mental health concerns, such as a psychiatric hospital. Some professional organizations also offer regional workshops at various sites around the country to help their members stay up to date on current issues within the profession. Sometimes these conferences take the form of interactive video presentations, a convenient way for large numbers of professionals to interact without meeting at a central site in the nation.

Course work at universities. Often, helping professionals discover, following their initial training, that they have just begun to learn. They choose to return to university-level schooling to take additional coursework, perhaps toward another advanced degree.

Professional meetings, such as local meetings of a chapter of a national organization. Often, someone locally who is knowledgeable about a particular area is the speaker on a timely local topic. At these meetings, helping professionals can also develop their own personal network and learn from each other. For the beginning professional, this contact is probably an undervalued method of continuing growth.

Making professional presentations. As one presents information to peers, often much personal growth occurs as one researches a topic and prepares a presentation. For this reason, making presentations sometimes carries professional development points, just as attending them does.

Journals, books, newsletters, videos, and other published sources

Most national professional organizations produce publications that help their members and other interested professionals attain new knowledge. Such publications include these:

Newsletters and other short publications. Typically, a newsletter will provide the helping professional with timely information: new developments in legislation, notices of conferences, employment opportunities, and other such information.

Journals that include accounts of new theories and effective practices. Because of a lengthy review process, journals do not always contain up-to-date information on very recent topics.

Books, monographs, conference proceedings, and other longer publications on topics of interest to helping professionals.

Videotapes and audiotapes of persons influential in the field.

Research and publication. Typically, a helping professional learns a great deal from conducting a research project or preparing a publication. Often, professional development credit is granted for publications.

Commercial programs and special-interest groups

Much up-to-date information is prepared and disseminated by commercial sources and special-interest groups. Examples of such materials include these:

Information on military careers, prepared and disseminated by all branches of the military. Recruiters sometimes arrange for helping professionals to tour military centers to acquire first-hand information on opportunities offered to inductees.

College, university, and proprietary school information, distributed by staff members of those institutions.

Pamphlets, visuals, and other information prepared by special-interest groups, such as career information disseminated by professional organizations, pamphlets prepared by family-planning agencies, substance-abuse information disseminated by various groups, and other such documents and visual programs.

Commercially prepared programs and presentations useful to helping professionals. These may include tests and inventories, computer programs, videotapes of interest to clients, and other such forms of information commercially marketed to helping professionals. Catalogs, samples, and professional seminars sponsored by those offering the materials may keep the professional informed, though caution must be advised to separate information from marketing claims.

Networking

Helping professionals are often each other's best sources of professional development. Experienced helpers, as well as novices, often turn to peers to verify conclusions, ask advice, or seek new sources of information. Personal interactions also seem to help the adult learner learn more easily and comfortably. A mentoring system, either formal or informal, can be the best way for a new professional to test out new ideas, check perceptions, and generally "learn the ropes" of the profession. Both mentor and mentee grow as they consider concerns and solutions.

Doubtless, the new helping professional will discover a variety of methods of continuing growth. Continued development is essential for the professional health of both the individual and the profession.

APPENDIX A

TFA Triangles

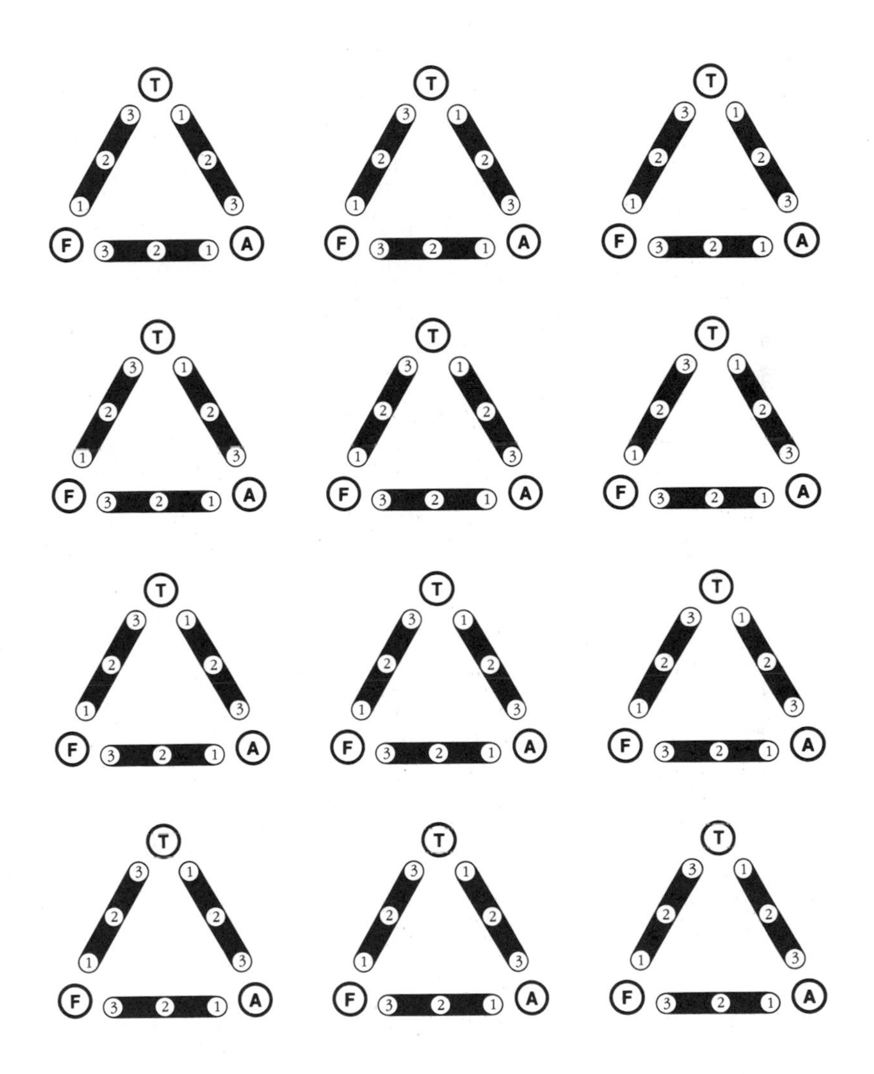

APPENDIX B

Sources of Ethical Standards and Professional Information

There are a number of sources of ethical standards that can be obtained from professional associations. However, one easily accessible source is *Issues and Ethics in the Helping Professions* (3rd edition) by Gerald Corey, Marianne Schneider Corey, and Patrick Callanan, published by Brooks/Cole. We strongly recommend this volume, which allows for convenient comparison of the following ethical standards:

- American Association for Counseling and Development
- American Association for Marriage and Family Therapy
- National Association of Social Workers
- National Federation of Societies for Clinical Social Work
- American Psychiatic Association
- American Psychological Association
- American School Counselor Association
- Commission on Rehabilitation Counselor Certification

Professional Associations

The following professional associations may be contacted for ethical standards as well as for membership materials and other information.

American Association for Counseling and Development
5999 Stevenson Avenue
Alexandria, VA 22304
(703) 823–9800

American Association for Marriage and Family Therapy
1717 K Street NW #407
Washington, DC 20006
(202) 429–1825

National Association of Social Workers
7981 Eastern Avenue
Silver Spring, MD 20910
(301) 565–0333

American Psychological Association
1200 17th Street, NW
Washington, DC 20036
(202) 955–7600

American Association of Pastoral Counseling
9504A Lee Highway
Fairfax, VA 22031
(703) 385–6967
This association has also completed a new *Code of Ethics*.

INDEX

A book such as this, oriented to teaching basic helping relationships and strategies, is but one part of a helper's initial training. We like to think of this as a springboard to continued lifelong learning and development. We wish you the best in your personal and professional development. We invite you to share with us your ideas, suggestions, and observations of what works for you, in terms of what we have written in this text.

Your school: ____________________

Instructor's name: ____________________

1. At what level (undergraduate/graduate) was the book used? ____________________

2. What kind of a program are you in? ____________________

____

____

3. What did you like *most* about *Helping Relationships and Strategies*? ____________________

____

____

4. What did you like *least* about this book? ____________________

____

____

5. Were all of the chapters of the book assigned for you to read? Please specify which were used.

____

6. What material do you think could be omitted in future editions? ____________________

____

____

____

7. In the space below or in a separate letter, please let us know your reactions to the book. Specific comments are most helpful. We'd really like to hear from you. Thanks, in advance, for your feedback!

Optional:

Your name: ______________________ Date: ______________________

May Brooks/Cole quote you, either in promotion for *Helping Relationships and Strategies* or in future publishing ventures?

Yes: ________ No: ________

Sincerely,
David E. Hutchins
Claire G. Cole

FOLD HERE

NO POSTAGE NECESSARY IF MAILED IN THE UNITED STATES

BUSINESS REPLY MAIL
FIRST CLASS PERMIT NO. 358 PACIFIC GROVE, CA

POSTAGE WILL BE PAID BY ADDRESSEE

ATT: *David E. Hutchins* and *Claire G. Cole*

Brooks/Cole Publishing Company
511 Forest Lodge Road
Pacific Grove, California 93950-9968

FOLD HERE